Electronic Troubleshooting and Servicing Techniques

By

J. A. Sam Wilson

and

Joseph A. Risse

©1998 by Howard W. Sams & Company

PROMPT© Publications is an imprint of Howard W. Sams & Company, A Bell Atlantic Company, 2647 Waterfront Parkway, E. Dr., Indianapolis, IN 46214-2041.

All rights reserved. No part of this book shall be reproduced, stored in a retrieval system, or transmitted by any means, electronic, mechanical, photocopying, recording, or otherwise, without written permission from the publisher. No patent liability is assumed with respect to the use of the information contained herein. While every precaution has been taken in the preparation of this book, the author, the publisher or seller assumes no responsibility for errors or omissions. Neither is any liability assumed for damages resulting from the use of information contained herein.

International Standard Book Number: 0-7906-1107-4
Library of Congress Catalog Card Number: 97-68175

Acquisitions Editor: Candace M. Hall
Editor: Loretta L. Leisure
Assistant Editors: Pat Brady, Natalie Harris
Typesetting: Loretta L. Leisure
Layout Design: Loretta L. Leisure
Cover Design: Debra Wilson
Graphics Conversion: Brian Drum, Bill Skinner, Margo Tuell, Terry Varvel

Trademark Acknowledgments:
All product illustrations, product names and logos are trademarks of their respective manufacturers. All terms in this book that are known or suspected to be trademarks or services have been appropriately capitalized. PROMPT© Publications, Howard W. Sams & Company, and Bell Atlantic cannot attest to the accuracy of this information. Use of an illustration, term or logo in this book should not be regarded as affecting the validity of any trademark or service mark.

PRINTED IN THE UNITED STATES OF AMERICA

9 8 7 6 5 4 3 2 1

To my wife Norma. She typed the complete manuscript. Also, to Mom and Dad, who claimed they were not responsible for any of the mistakes I have made since my twelfth birthday.

J. A. SAM WILSON

As co-author, I would like to dedicate my work in writing this book to my wife Anne, and my six grand-children: Ryan J. Risse, Matthew J. Risse, Stephen V. Armitage, Jane E. Armitage, Kathleen E. Risse and Daniel J. Risse.

JOSEPH A. RISSE

Table Of Contents

Chapter 1
The Basis for Testing and Troubleshooting 1
   Overview 1
   Objectives 1
   ANALOG 2
   Where Do You Start? 2
   Symptoms - An Introduction 2
   The Statistical Approach 3
   Troubleshooting with Test Equipment 3
   Qualitative and Quantitative Analog
      Circuit Measurements 9
   Know Your Test Equipment 12
   When Measurements Won't Help 13
   Measurement Efficiency 14
   DIGITAL 14
      Symbolism 15
      Glitches 15

IDEAS YOU SHOULD CARRY OVER
FROM YOUR WORK IN ELECTRONICS 16
IDEAS YOU SHOULD NOT CARRY OVER
FROM YOUR WORK IN ELECTRONICS 18
THE BASIC GATES 20
THE AND GATE 21
THE OR GATE 22
THE NOT GATE 23
THE NAND GATE 23
THE NOR GATE 26
THE EXCLUSIVE OR GATE 27
LOGIC FAMILIES 27
TTL (TRANSISTOR-TRANSISTOR LOGIC) 29
ECL (EMITTER-COUPLED LOGIC) 29
CMOS (COMPLEMENTARY METAL OXIDE SEMICONDUCTOR) 29
RTL (RESISTOR-TRANSISTOR LOGIC) 30
SUMMARY 30
CHAPTER 1 QUIZ 31

CHAPTER 2
TROUBLESHOOTING BASIC ANALOG CIRCUITS 37

OVERVIEW 37
DC VOLTAGES ON AMPLIFYING DEVICES 38
MODELS 45
A RESISTOR MODEL 48
CLASSES OF AMPLIFIERS 48
CONFIGURATIONS 50
DC SUPPLY CONNECTIONS 51
METHODS OF OBTAINING BIAS 53
COUPLED CIRCUITS 57
LEVEL SHIFTING 59
DIFFERENTIAL AMPLIFIERS 60
POWER AMPLIFIERS 61
THE RECEIVER AS AN ELECTRONIC SYSTEM 64
FINDING TROUBLE IN A DEAD RECEIVER 70
OPERATIONAL AMPLIFIERS 71
TESTS FOR TRANSISTORS 72

CIRCUITS THAT COULD BE REPRESENTED BY LOGIC GATES 76
SUMMARY OF LINEAR CIRCUITS 76
CHAPTER 2 QUIZ 76

CHAPTER 3
TROUBLESHOOTING WITH METERS AND LOGIC PROBES 83
OVERVIEW 83
OBJECTIVES 83
ANALOG 84
ABOUT THE METER 85
MEASURING AC CURRENT 87
OHMS PER VOLT 88
METER LOADING 89
CALIBRATION 90
TAUT-BAND AND JEWELED ANALOG METERS 93
ACCURACY OF VOLTMETER MEASUREMENT 93
METER PROBES 94
BATTERY TERMINAL VOLTAGE 95
ADDITIONAL INFORMATION ON OHMMETER TESTING 96
TESTING CAPACITORS WITH A VOM 99
DIODE TEST WITH AN OHMMETER 101
ESTIMATING VOLTAGES 101
AMPLIFIER VOLTMETER TEST 102
  SHORT-CIRCUIT VOLTMETER TEST 104
DIGITAL 105
INTRODUCTION TO COMBINATIONAL LOGIC 105
CIRCUITS MADE OF INDIVIDUAL GATES 107
WRITING OUTPUT EQUATIONS FOR COMBINED LOGIC GATES 109
FINDING THE OUTPUT LOGIC LEVEL OF COMBINED GATES 110
  TIMING DIAGRAMS 110
ANALOG SUMMARY 112
DIGITAL SUMMARY 113
CHAPTER 3 QUIZ 113

Chapter 4
Troubleshooting with an Oscilloscope 117
Overview 117
Objectives 118
Comparison of Oscilloscope Types 119
ANALOG 122
Oscilloscope Bandwidth 122
Delay 124
Frequency vs. Time Domain 125
Oscilloscope Accessories 126
Voltage Calibration 126
Frequency Calibration 127
Chop and Alternate 128
Low-Capacity Probes 129
Parallax in Oscilloscope Measurements 129
Measuring Current 130
Determining Average and RMS Value 130
Measuring/Evaluating Components With an Oscilloscope 132
Amplifier Troubleshooting with an Oscilloscope 132
Frequency Distortion 132
Linear Distortion 134
The Amplifier Lissajous Test 137
Intermodulation Distortion 138
DIGITAL 140
The RS Flip-Flop 140
Clock Circuits 145
Combined Circuits on an Integrated Circuit Chip 146
JK Flip-Flops 147
Schmitt Trigger 149
Analog Summary 149
Digital Summary 149
Chapter 4 Quiz 150

Chapter 5
Signal Injection and Signal Tracing 153
Overview 153
Objectives 153

SYSTEM TROUBLESHOOTING 154
    CUTTING THE TIME REQUIRED FOR SIGNAL
        INJECTION OR SIGNAL TRACING 154
    USING A RADIO AS A SIGNAL INJECTOR 156
    SIGNAL INJECTION 157
    SIGNAL TRACING 159
    SIGNAL INJECTION WITH THE AMPLIFIER SIGNAL 160
    USING THE PROPER PROBE 161
    USING THE FREQUENCY DOMAIN DISPLAY 161
    USING THE SWEEP TECHNIQUE 162
THE ROLE OF AFC AND AGC 164
NOISE GENERATORS 165
DIGITAL 166
    SERIAL AND PARALLEL TRANSMISSION OF DATA 166
SUMMARY 169
CHAPTER 5 QUIZ 169

CHAPTER 6
SYMPTOM ANALYSIS, DIAGNOSTICS AND STATISTICAL METHODS 175
OVERVIEW 175
OBJECTIVES 176
SYMPTOMS 176
    OPEN THE CABINET 177
    THE DIAGNOSTIC 178
STATISTICAL ANALYSIS 180
SUMMARY 182
CHAPTER 6 QUIZ 183

CHAPTER 7
SERVICING CLOSED-LOOP CIRCUITS 185
OVERVIEW 185
OBJECTIVES 186
CLOSED-LOOP CIRCUITS FOR VOLTAGE CONTROL 186
    FREQUENCY DOMAIN 190
    THE AVC/AGC BIAS VOLTAGE 191
ANALOG (LINEAR) VOLTAGE REGULATORS 193

SWITCHING REGULATORS 195
CURRENT REGULATION 196
FEEDBACK CIRCUITS THAT CONTROL/ESTABLISH FREQUENCY 197
AUTOMATIC FREQUENCY CONTROLS (AFC) 199
THE PHASE-LOCKED LOOP 200
REVIEW OF SOME BASIC CIRCUITS USED IN CLOSED LOOPS 203
OP AMP LOGIC COMPARATOR 203
SUMMARY 204
CHAPTER 7 QUIZ 205

CHAPTER 8
HUNTING FOR THE CAUSES OF NOISE AND INTERMITTENTS 209
OVERVIEW 209
OBJECTIVES 209
ANALOG 210
RESISTOR NOISE 210
WHAT IS BOLTZMANN'S CONSTANT? 213
AN EXAMPLE OF NOISE 215
EXTERNAL NOISE 219
AN EXAMPLE OF A NOISE PROBLEM 219
INTERMITTENTS 220
DIGITAL 221
CLOCK CIRCUITS 221
COMBINED CIRCUITS ON AN INTEGRATED CIRCUIT CHIP 222
JK FLIP-FLOPS 223
SCHMITT TRIGGER 226
TYPES OF NOISES AND THEIR CAUSES 226
SUMMARY 227
CHAPTER 8 QUIZ 227

CHAPTER 9
SERVICING DIGITAL LOGIC AND MICROPROCESSOR EQUIPMENT 231
OVERVIEW 231
OBJECTIVES 232
POWER SOURCES AND PROPAGATION DELAY 232

Three-State Devices	233
Knowledge as a Valuable Troubleshooting Aid	238
Test Equipment Useful in Troubleshooting Logic and Microprocessor Circuits	238
Logic Pulsers	239
Frequency Counters	240
Microprocessors	241
An Example of a Microprocessor System	241
Troubleshooting a Microprocessor System	245
Summary	246
Chapter 9 Quiz	246

Chapter 10
Soldering Techniques for Repair and Replacement 249

Overview	249
Objectives	250
Soldering Techniques	250
The Solder	250
Some Important Dont's	253
Soldering	254
Inspecting the Solder Connection	256
The Soldering Iron	256
Unsoldering Connections	257
Additional Soldering Tips	258
Summary	260
Chapter 10 Quiz	260

Chapter 11
Low-Cost Homemade Testing Devices 263

Overview	263
Objectives	264
Nonpolarized Electrolytic Capacitors	264
Voltages to Check Calibration	265
Using a VOM for a High-Voltage Measurement	267
A Basic Logic Probe	268
A Basic Noise Generator for Use in Signal Tracing	269

MEASURING AC CURRENT WITH A VOM 270
A RESISTOR SUBSTITUTION DEVICE 271
DETECTOR PROBE 272
A SIMPLE IN-SITU (ON-SITE) TRANSISTOR TESTER 274

APPENDIX A
   EXPERIMENTER'S LOGIC PROBE 277
APPENDIX B
   BINARY COUNTING 279

APPENDIX C
   BOOLEAN ALGEBRA 281

APPENDIX D
   QUIZ ANSWERS 283
      CHAPTER 1 283
      CHAPTER 2 286
      CHAPTER 3 288
      CHAPTER 4 289
      CHAPTER 5 290
      CHAPTER 6 291
      CHAPTER 7 292
      CHPATER 8 293
      CHAPTER 9 294
      CHAPTER 10 294

PREFACE

I once wrote a test procedure for fixing an industrial electronic system. I started by warning the technicians not to operate the equipment with the safety interlock defeated. (To do so would subject the technician's body to a very high dose of radiation.) I was strongly criticized by an editor for starting on a negative note.

To this day I don't know how I would write the procedure without first warning the technicians about defeating that interlock. Maybe I was supposed to say, "Go ahead, defeat the interlock and see what you get."

Fortunately, there are no life-threatening procedures in this book. However, in addition to explaining the purposes of the book I also want to say what the book is NOT intended to do.

My writing experience aside, I'll start with something the book does not do. This is NOT a book on how to repair a VCR, or a television receiver, or a radio, or any other specific piece of equipment. Rather it IS a compendium of troubleshooting tests, measurement procedures, and servicing techniques. There is also some information on how to make repairs and replace components.

For the purposes of this book, the hierarchy of electronics - from the simplest to the most complicated - is given here:

Component - Separate parts that combine to make an electronic circuit: resistors, capacitors, and inductors. Also, active components (tubes, transistors, and FETs) are included. Digital components (logic gates and flip flops) are also included in some chapters.

Circuit - A combination of components used to perform a specific function: oscillators and power amplifiers.

System - A combination of circuits: radios and computers.

In the book I discuss tests, measurements, and troubleshooting for all of those categories. For components and circuits there is a wide variety of examples. However, depending on level of schooling and experience, there are very few systems that all readers would be acquainted with.

If I used a television receiver, or a VCR, or an industrial control system as an example, there would be readers who are not comfortable with the subject. So, I chose a radio for discussing systems. For those who would like to review the basic radio system, I have included a short discussion of a typical radio in Chapter 1. The purpose of every component in that radio is included. If you know what a component is supposed to do, you will be better able to recognize the problem when it is not doing its job. (This is an opinion - not a hard and fast rule.)

Because of the wide variety of problems that can occur in electronic equipment, I seriously doubt if it is possible to have a universal, fits-all troubleshooting procedure. I know that many authors have tried to formalize a standard approach and most of those approaches are well thought out. I have never been able to make that approach work for myself.

However, I know that some of my readers would be disappointed if I didn't at least try, so here is my recommended procedure:

> Learn as much as you can about electronics. I presume you know that subject, but a little review is always a good idea. Chapter 2 reviews some of the basics as they apply to troubleshooting.
>
> Look for obvious possibilities.

If there are no obvious faults, measure the power supply voltage. If it is not the correct value, don't go any further until you find out why.

Use symptoms whenever possible to zero in on the section at fault. This method is discussed in Chapter 6. If symptoms don't get you to the trouble region, use a signal tracing or signal injection method, as discussed in Chapter 5.

If those methods fail, use diagnostics as discussed in Chapter 6.

Once you have zeroed in on the troubled circuits, use the methods discussed in Chapter 3 and 4 to locate the defective component.

Tough problems (like closed loops, distortion, noise and intermittents) often require special techniques like those discussed in Chapters 7 and 8.

Make efficient use of your test equipment, as discussed throughout the book, especially in Chapters 3 and 4. Specialized equipment that you build yourself can be very useful for locating defective components. Chapter 11 and Appendix A describe a few useful testers you can build.

Digital and microprocessor circuits require special equipment and a few special techniques. They are discussed throughout the book.

When looking for faulty components it is a good idea to keep in mind the fact that troubleshooting often involves these steps:

> Make a measurement.
>
> Compare the measured value with what you are supposed to get.
>
> If the measured values do not match what you are supposed to get, find out why. This book describes specific measurements used for locating faulty components in circuits, and systems.

When you have located the defective component, do a professional job of replacing it. Some newer techniques are needed for surface mounts. Chapter 10 discusses the replacement of components.

I realize that this procedure has more than two or three steps, but it has as few as I can make it.

I believe your best troubleshooting help will always be your knowledge of how electronic things work. I have assumed you can use a voltmeter, an oscilloscope, a signal generator and a frequency counter for their basic purposes. For example, it is assumed that you can use a voltmeter (or VOM) to measure a voltage. Use of test instruments for basic purposes is *not* a subject for this book.

The book is organized by techniques rather than by an organized troubleshooting procedure. In some cases those techniques overlap chapters. For example, should the signal injection technique be combined with the signal generator applications or with the system analysis discussions?

After spending hours debating questions like that, I decided the most important thing is to get the techniques into the book. (I'm sure the readers can put them in their own favorite categories.) However, you will see a general drift of ideas from component-to-circuit-to-system analysis whenever possible.

The selection of test equipment depends on the amount of service and the type of service done. As with universal troubleshooting rules - which probably don't exist - it is likely that there is no perfect list of test equipment. So as not to be a disappointment to the reader, I will make a valiant effort to make a list anyway:

Analog VOM

Digital VOM

Dual-trace, triggered sweep oscilloscope (50 MHz or higher preferred)

ESR (Equivalent Series Resistance) meter (for electrolytic capacitors)

Function generator with a VCO (Voltage Controlled Oscillator) sweep provision

Logic probe and logic pulser

Low-distortion sine wave generator (audio range)

Regulated DC power supplies +3V, +3.6V, +5V, ±12V

Transistor testers

DC power supplies (stiffly regulated) +3V, +3.6V, +5V, ±12V

Each chapter in this book starts with an overview of the subject and a list of specific objectives. Following that, the subject matter is discussed in detail.

Quizes at the end of each chapter give the reader a chance to demonstrate an understanding of the subject matter. For those who use the book for self-study, the answers are given to all questions in Appendix D.

There are some questions on subjects not covered in the text. They are given for a review of basic theory. If you cannot answer any of those questions it might be a good idea to review the related subject in your basic textbook.

I gathered many of the ideas in this book from technicians who attended my lectures. Some technicians have written details of their favorite techniques in their letters to me. Technicians in Kansas and Wisconsin have made special efforts to help. In some cases, they are ideas that I put to use in my own troubleshooting experience. If I included anyone's name I would have to include everyone's name. Frankly, I can't remember all of those names. Those of you who have helped - you know who you are - have my deepest gratitude.

I am also indebted to company training brochures and other training programs.

I have learned much from the students who have passed through my theory and lab classes. Many were working in electronics and attending school at the same time.

Last, but certainly not least, much appreciation must be given to my wife Norma who helped all the way. She is more than a computer operator. In my company I am the president and she is everybody else.

J. A. Sam Wilson

It is with great pleasure that I am accepting this opportunity to co-author this book with my long-time friend, Sam Wilson. I would also like to express my appreciation at this time to someone who has provided great editorial assistance, Natalie Harris, Project Manager at Prompt® Publications.

This book covers the description and use of both analog and digital electronic instruments. These days, an electronic technician must be familiar with both types. Since the earliest days of electronics, an increasing dedication has been required to keep pace with advancements.

It is fortunate that many people find the study of electronics challenging and interesting. It requires more study and more time than most other fields of endeavor, but the rewards are proportionate. May each of you as a reader, a student, technician, hobbyist or engineer enjoy great success!

Joseph A. Risse

CHAPTER ONE

THE BASIS FOR TESTING AND TROUBLESHOOTING

OVERVIEW

There is no such thing as a perfect troubleshooting procedure that works for every occasion and for everyone. If there was such a thing, every good technician in the country would be using it. Instead, technicians have their own favorite approaches and favorite test procedures that have been developed over years of making a living in the servicing business. However, certain basic procedures are similar in all testing procedures, and they are the subject of this chapter. Each chapter is divided into two sections: ANALOG and DIGITAL.

OBJECTIVES

How much can you rely on symptoms as a troubleshooting aid?

What are some key words to listen for when an owner or user is describing symptoms of faulty equipment?

How can a milliammeter be used to check a power amplifier?

What are the statistical chances of a component being bad?

How is a resistor sometimes used as a fuse?

What are some of the factors that must be considered when analyzing a qualitative measurement?

How do qualitative and quantitative tests differ?

What are the basic logic gates that serve as a foundation for digital troubleshooting?

How is a logic probe used for troubleshooting digital circuits?

What are the logic families?

ANALOG

WHERE DO YOU START?

A good way to start troubleshooting is to give everything a close visual inspection. That can be done very quickly. It is a good idea to look for such obvious problems as burned spots and places where a high voltage arc has occurred. Also, inspect the fuses or circuit breakers.

Be sure to look at the screws holding the back and chassis. Marked-up screws indicate that an unprofessional person may have been tinkering with the system. That, in turn, may indicate a major troubleshooting problem.

Make sure your work indicates that high-quality professional servicing has been done. A sloppy repair job is like a signature you have left for other technicians to see. Things like a poor soldering job, damaged screw heads, frayed wires, scratches on the cabinet, etc., are an indication that an unprofessional person has been at work. Take time to do it right!

SYMPTOMS – AN INTRODUCTION

Symptoms may be useful for locating the general area of the problem. The symptoms are sometimes described by the equipment owner or user. In an industrial electronics plant, it may be the foreman of the division who uses the equipment. In consumer electronic equipment, the description is often from the customer who owns the equipment.

Experienced technicians have learned to be wary of users' and owners' descriptions of symptoms. Because those people are usually not electronic experts, they are likely to describe a problem in terms of things they are already familiar with. Here are some key things to listen for: fire, smoke, overheating or the smell of something burning.

The most reliable symptom analysis is given by the technician after energizing the equipment. Technicians know that certain symptoms in a system usually mean that a certain component has failed. That failure has often been observed when previous work has been done on the same system.

Some technicians keep track of recurring problems in a personal notebook. In the case of consumer electronic systems, the symptoms and cures are available on computer databases.

Symptoms are indeed a valuable guide, but it is not a good idea to base a complete troubleshooting procedure on symptoms alone. For example, distortion in a radio's output sound can be caused by different problems such as a tear on a speaker cone, an overdriven amplifier or low terminal voltage of an aging battery. Distortion can also be caused by a transistor that is no longer able to do its job. More will be said about symptoms in Chapter 6.

THE STATISTICAL APPROACH

Although it is not usually discussed, experienced technicians often use a type of statistical analysis when troubleshooting. In the case of sound distortion just described, there are several possible faults for the same symptom. From experience, the technician knows that a battery will fail more often than a transistor. So, the battery would be checked and replaced first. *Table 1-1* lists component failures in their order of probable occurrence.

Because it is usually easier to replace a battery than a transistor, ease of replacement is also a factor. Who doesn't take the easy path first? So, if there are two equal choices, you will likely try the easier repair first. However, given equal replacement difficulties it is best to start by testing the component that is most likely to fail.

TROUBLESHOOTING WITH TEST EQUIPMENT

Having done the visual inspection, you are ready to take a serious look at the faulty equipment using your test equipment. The first test equipment measurement should be the same for every troubleshooting problem: measure the power supply voltage when the system is energized.

The Basis for Testing and Troubleshooting

Mechanical and electromechanical devices such as relays and switches. Also plugs and sockets, especially if used often.

Components that get hot in their normal operation, like power amplifiers and rectifiers.

Electrolytic capacitors. Especially very small versions and those subjected to high voltage.

Active devices like transistors and SCRs.

Passive devices like resistors and capacitors.

Table 1-1.

You can be sure that nothing is going to work right in the system if the power supply voltage is not the correct value. A low-power supply voltage can cause the failure of a circuit that is not located anywhere near the supply. It is best to measure the power supply voltage while it is delivering current to the system. In battery-operated equipment, the system should be energized when the battery voltage is being measured.

Remember that an incorrect power supply voltage can damage a digital circuit. In many cases a stiffly-regulated power supply is used for digital circuits. If you encounter a wrong voltage in a digital circuit shut it down immediately. Then use your bench supply to energize the circuit and look for possible damage caused by the incorrect voltage.

Figure 1-1 shows two examples of basic power supplies for analog systems. *Figure 1-1a* is a basic half-wave rectifier. *Figure 1-1b* shows a typical battery supply. The half-wave rectifier circuit of *Figure 1-1a* will be used as the first example.

If there is no output voltage from the supply, check the obvious possibilities such as a defective switch, blown fuse or open diode. Statistically, the fuse is the most likely to be the failed component. Be careful. High voltages may exist.

There is a very definite advantage to using an autoranging digital voltmeter in this type of measurement. With an analog meter, when you are not sure what value of voltage or current you are going to be measuring, you have to start with the highest scale and work down. With a digital (autoranging) meter, the correct scale and voltage polarity for measurement will automatically be selected.

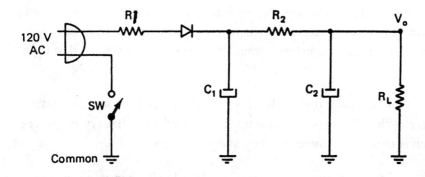

Figure 1-1a. A battery with a decoupling filter.

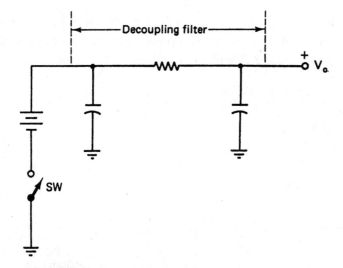

Figure 1-1b. A half-wave rectifier.

Figure 1-2 shows quick ways to check a switch and a fuse. If you are using an analog meter, be sure to set the meter scale to a value that is high enough to accommodate the full circuit voltage or current.

For the connection in *Figure 1-2a* the meter display should show a high voltage when the switch is open, and no voltage when the switch closed. Even though the switch makes a clicking noise when it is operated, a faulty switch may be permanently open or closed.

The connections of *Figure 1-2b* or *Figure 1-2c* can be used to check the fuse or switch. If the fuse is open, you will measure the full DC current. You should know the amount of current normally drawn by the load. If not, you can set the analog current meter to the maximum current scale, then decrease the scale until you get a deflection.

There is an advantage to the method of using a current meter in the tests just described. The current measurement tells you if a replacement fuse can be used. If the current is excessive, that is, higher than the defective fuse rating, start looking for the trouble right away.

The meter in *Figure 1-2c* should measure the full battery current under load *when the switch is open*. (Remember, the term "load" refers to the current supplied by a battery or power supply.) When the switch is closed, the meter reading should be zero.

In the battery supply, a defective switch that is permanently closed will often result in a dead battery because the equipment has likely been on for a long time.

Suppose the power supply is delivering current to an amplifying system. The power amplifiers will require most of the power supply current. If a power amplifier is open, or cut off, the meter display in *Figure 1-2b* and *Figure 1-2c* will show a current that is too low.

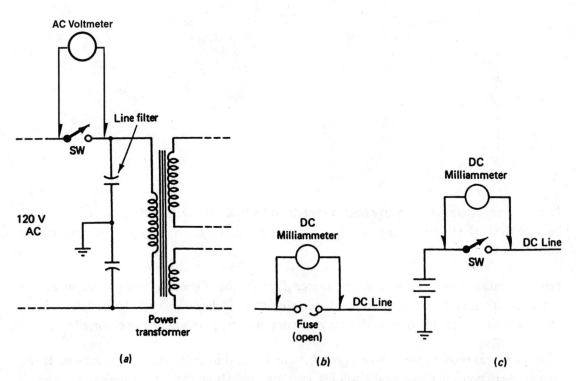

Figure 1-2. Methods of checking the on/off switch. a) The switch is in the primary circuit of the power transformer. b) A fuse is like a switch. When the fuse is open, the DC milliameter completes the circuit. c) In this case, the DC milliameter is connected across the open switch to measure the total current drawn by the circuit.

A common problem with bipolar transistor power amplifiers is an emitter-to-collector short circuit. This will result in a current reading that is too high as indicated by the DC milliammeter connection.

If the current is zero for the open and closed switch positions, you can easily tell if it is due to a defective switch or some other power supply defect. The DC milliammeter will complete the circuit and you will measure an output current unless the switch is stuck in the closed position.

There doesn't appear to be a fuse in the circuit of *Figure 1-1a*. However, the surge-limiting resistor (R_1) may have a low enough wattage rating to protect the circuit. In other words, the power rating of that resistor may be purposely chosen so that the resistor acts like a fuse in case of an excessive current.

Measure across R_1 with a voltmeter (*not* a milliammeter). A high voltage indicates that the resistor may be open. If you connect a DC voltmeter across the output of the supply in *Figure 1-1a*, remember that the output voltage (V_o) may be as high as 160V or more. The 120V AC input is an RMS value, but the capacitors charge to the peak value of that voltage.

Leaky filter capacitors can cause an excessive current through the diode. In the case of a leaky C_2, excessive current may flow through filter resistor R_2 lowering the supply output voltage. If C_1 is leaky, then the diode may be destroyed or R_1 may be open. If the diode is burned out or R_1 is open, check the electrolytic capacitors *before* making a diode replacement.

Excessive ripple in the power supply output may be caused by a defective filter capacitor. Technicians often use the bridging method to check for this possibility by connecting a good capacitor across C_1 and then across C_2. Another method of checking the filter capacitors is to look for excessive ripple with an oscilloscope by using the AC input on the scope.

A unique way of checking electrolytic capacitors is shown in *Figure 1-3a*. The AC voltmeter is connected in series with the electrolytic capacitor. In this application the voltmeter is being used as a current meter. Excessive AC current indicates a defective capacitor. A disadvantage of this method is that you have to take time to disconnect the capacitors. Also, you have to know (from experience) the allowable AC current since electrolytic capacitors are notoriously leaky; that is, they act as if they have a parallel resistor to allow current flow.

You should not try to determine the condition of an electrolytic capacitor by using an ohmmeter, see *Figure 1-3b*. Most ohmmeters measure resistance by supplying a low voltage to the device being measured. That low voltage, which is often only 1.5 volts, does not accurately measure the leakage resistance of an electrolytic capacitor.

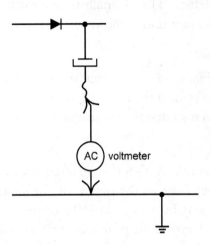

Figure 1-3a. Special method of testing electrolytic capacitors and power supplies.

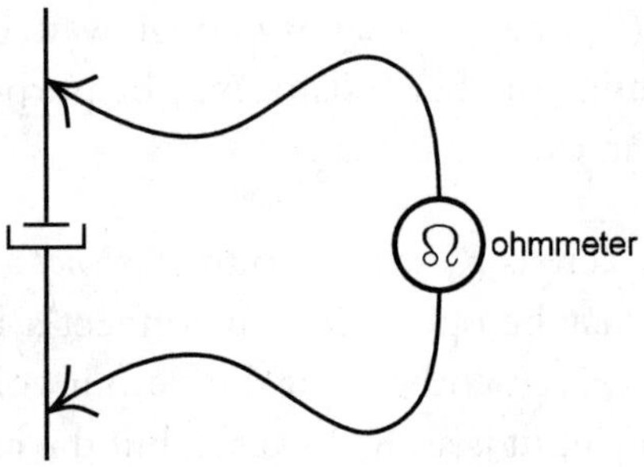

*Figure 1-3b. This is **not** a valid test of an electrolytic capacitor.*

An accurate method of measuring the condition of an electrolytic capacitor is to use an ESR (Equivalent Series Resistance) meter. Some digital multimeters have a built-in ESR measuring capability. An ESR measurement takes into consideration the series and leakage resistance of the capacitor.

If there is a low power supply output voltage, be sure to check the AC power line voltage. Defective rectifiers or filter capacitors are highly-probable causes of low supply voltage.

It is not rare that a low power supply voltage is being caused by a defective circuit in the system being supplied. For example, a short circuit to ground in an amplifier can cause the power supply to be excessively loaded. That in turn can cause a fuse to blow, a diode to burn out or a low output voltage.

The troubleshooting techniques described here will work for other types of unregulated supplies. More will be said later about power supply troubleshooting.

QUALITATIVE AND QUANTITATIVE ANALOG CIRCUIT MEASUREMENTS

Assuming that the power supply voltage is OK, additional measurements are in order. It is important to understand the types and limits of measurements.

There are two major kinds of measurements: *qualitative* and *quantitative*. With a qualitative measurement, you must make a judgment. An example is shown in *Figure 1-4*. The desired pattern on an oscilloscope screen is sometimes provided by the manufacturer of the equipment that is being serviced. You use your experience and knowledge of the system to determine whether the displayed pattern is sufficiently close to the manufacturer's specified pattern.

Qualitative measurements are very difficult to teach and learn. It would be helpful if the technician would spend some time making qualitative measurements with equipment that is working properly. In a short while the required experience can be gained for interpreting results. Important questions to ask when making a qualitative judgment include:

> Can the symptoms of the defective system be produced by the variation in patterns?
>
> Is the maximum height (peak-to-peak value) on an oscilloscope display an important factor?
>
> Is the width (time duration) of the scope pattern an important factor?
>
> Is it possible to make any adjustments to get the waveform obtained closer to the desired pattern?
>
> Is it possible that a defective component is causing the variation?

Manufacturer's specified scope pattern

Scope pattern obtained during troubleshooting

Figure 1-4. A qualitative judgment is required to interpret the pattern.

Quantitative measurements provide numbers. A voltmeter measurement is quantitative. If the manufacturer's specification says that a certain voltage must not be less than 30V, and the voltmeter displays a voltage of at least 30, then the measurement has given an indication that there is no trouble at that point.

In some cases it is necessary to make judgments with quantitative measurements. If the schematic diagram indicates that a voltage should be 10V and you measure 8.6V, you will have to determine the effect of that lower voltage. A voltage of 8.6V would not be acceptable for a stiffly-regulated 10V supply. However, it might be OK as the collector voltage of a voltage amplifier.

The power supply voltage for a digital circuit is often stiffly regulated. As mentioned before, if the power supply voltage for a digital circuit is out of tolerance, shut the system down immediately. Quickly check for an overheated IC (integrated circuit). That could be your first replacement.

Some measurements are not exactly in either the qualitative or quantitative category. They can be both. A good example is the square wave test. It is illustrated in *Figure 1-5*. A square wave is applied to the input of an amplifier and the output square wave is observed on an oscilloscope. For a qualitative judgment, the technician observes the shape of the output square wave. Knowledge of possible problems, as indicated by the square wave test, are put into play. A quantitative judgment involves amplitude and frequency.

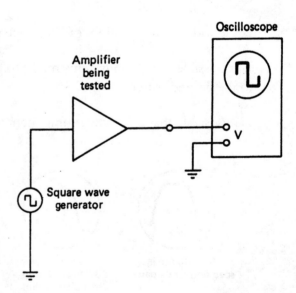

Figure 1-5. A square wave test is usually qualitative, but under certain conditions it can be used as a quantitative test. For a low-frequency amplifier, such as an audio amplifier, use a 1000 Hz square wave.

Figure 1-6 shows some of the possibilities. A technician has to make the decision if a slightly improper square wave is still acceptable as far as the amplifier is concerned. If you use a square wave generator with a variable frequency output and adjust the frequency high enough, most amplifiers will show some kind of high-frequency distortion. So, the frequency at which the square wave generator is operated is a very important factor.

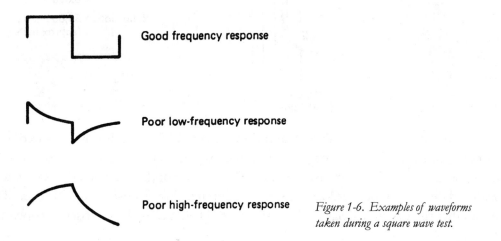

Figure 1-6. Examples of waveforms taken during a square wave test.

Some textbooks say that the square wave test can be used as an indication of bandwidth in a quantitative measurement. One suggested procedure is to raise the frequency of the input square wave until the output becomes slightly distorted. Then the frequency of the square wave (at which the distortion first occurs) is multiplied by 7 or 9, corresponding to the seventh or ninth harmonic. That (supposedly) gives the bandwidth of the amplifier.

Another quantitative evaluation can be made by measuring the rise time of the square wave on a triggered-sweep oscilloscope. Then using the following equation:

$$\text{Bandwidth} = \frac{0.35}{\text{Rise time}}$$

Here the rise time is expressed in seconds and the bandwidth is expressed in megahertz. For a low-frequency circuit use 1000 Hz for the test.

Figure 1-7 shows how the rise time of any pulse or square wave is measured. Note that the rise time is defined as the amount of time that it takes the wave to go from 10 to 90 percent of the maximum amplitude. You can use an X10 sweep expander (available on many scopes) to make it easier to read the rise time. Be sure to divide the displayed rise time by 10 to get the real value.

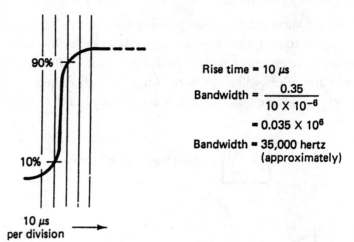

Figure 1-7. *Definition of the term rise time as it applies to the leading edge of a step function.*

You have to be very careful in making that measurement. If the rise time is measured on an oscilloscope, and the scope cannot pass the square wave, then the test is useless. For example, suppose the rise time is measured to be 180 microseconds, but when the square wave is introduced directly to the oscilloscope's vertical input it shows a rise time of 180 µs. That test shows that the rise time delay in the test is most likely caused by the oscilloscope.

So, you are using square wave analysis as a qualitative test if you are making a judgment on the shape of the output waveform. You are using it as a quantitative test if you use the rise time to determine the bandwidth.

KNOW YOUR TEST EQUIPMENT

You must know the capabilities, limitations, and applications of the equipment you are using. Those things are best determined by carefully reading the manufacturer's equipment specifications. Say you are looking for glitches (undesired spike voltages). If you are using an oscilloscope that has a bandwidth of 20 megahertz it is doubtful you will see any glitches and thus conclude, incorrectly, that there are none. The real problem is that the bandwidth of the oscilloscope is too narrow compared to what is required for displaying the glitch that you are missing. You are using a piece of equipment that is not able to perform the measurement you are trying to make.

Another example is shown in *Figure 1-8*. A volt-ohm-milliammeter (VOM) with an input resistance of 100,000 ohms cannot be used reliably to measure a voltage drop across a 5-megohm resistor. The relatively low resistance of the voltmeter is in parallel with the 5 megohms, and their combined resistance is less than 100,000 ohms. That greatly reduces the voltage drop you are trying to measure.

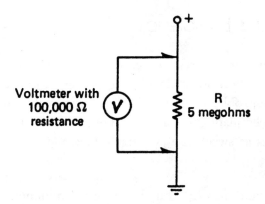

Figure 1-8. Meter loading occurs when the resistance of the meter is much lower than the resistance across which the meter is connected.

Of course, if you know the resistance of the voltmeter, you can compute the actual voltage on the basis of the measurement taken. A much better idea, though, is to use a high-impedance digital voltmeter (or an oscilloscope) to make the measurement.

The manufacturer's specifications for an instrument contain information that a technician should know and understand when using it to make measurements.

WHEN MEASUREMENTS WON'T HELP

Accurate measurements are very important in troubleshooting, but there are some types of troubleshooting procedures where meters and oscilloscopes may not provide useful information. As an example, a high-gain amplifier that is oscillating can produce an undesired signal in other stages. The oscillation may be originating in one amplifier stage, but you may find it in all of the following stages. An intermittent circuit is another case where making a measurement may not help.

So, to put measurements in their proper perspective, it should be said that they are useful in most troubleshooting procedures. In the cases where a measurement doesn't help, the technician must resort to other test procedures.

A troubleshooting procedure for intermittents and oscillations, known as the *shotgun* method, must be mentioned here. This method is distasteful to many technicians. Essentially it requires that once everything else fails, it may be necessary to replace each part, one at a time, until the problem is solved.

For most technicians that is a last resort. However, there are a few times when none of the tricks of the trade will work. There are some cases like complicated impedance-coupling networks where the manufacturer actually suggests the shotgun approach rather than spending many days on the problem.

Measurement Efficiency

It is not productive to make a measurement that requires five pieces of test equipment to locate a problem that could just as easily have been located with a voltmeter or an oscilloscope. As a rule, the more complicated the problem the more likely a basic measurement will not be useful. But, you have to start by eliminating the obvious. This is done by making basic measurements before such techniques as sweep analysis, signal injection and signal tracing are undertaken.

Measurements for troubleshooting should be done efficiently by using the minimum test equipment necessary to locate the problem. This will increase your speed in locating a trouble source.

DIGITAL

The first step in learning digital troubleshooting is to learn the components that are available, and what each component does in a circuit. There are two reasons this should be done. First, unless you know what a part is supposed to do, you can't tell if it is doing its job. Second, with the number of basic components available there is an almost unlimited variety of circuits and systems possible. It is not possible to learn all of them at one time, but if you know your basics you can understand the circuits or systems that you have to work with.

The second step in learning digital troubleshooting is to study basic circuits that are made from the components. If you were studying television, you might first learn what transistors, resistors, capacitors and inductors do in a circuit. Then, you would learn some basic circuits like oscillators and amplifiers. To learn troubleshooting in digital systems you should start with basic logic components (such as AND, OR and NOT gates) then learn some basic combinational logic circuits (like flip-flops and enables).

The third step in learning digital troubleshooting is to study how the basic circuits are combined into complete systems. In studying television, you would learn methods of coupling amplifiers, types of mixers for combining oscillator and RF circuits, etc. In learning digital systems you would study counters, an example of logic used in a tuner, and other applications.

Just knowing how the components, circuits and systems work will not make you a troubleshooting expert. You must also know how to conduct tests and measurements; and you must know symptoms. You cannot become a technician by learning only symptoms, but a n understanding of symptoms should be a part of your total knowledge.

As you get more involved in digital troubleshooting you will want to add some specialized information to the material given in this book. For example, you may want to learn about Boolean algebra, Venn diagrams, minterms, maxterms, binary arithmetic, etc. You will have to study those subjects later in order to be an expert in troubleshooting logic circuits.

SYMBOLISM

Unfortunately, there are a number of different types of schematic symbols being used for logic circuits. By far, the most popular are the MIL symbols.

If you do not know which kind of symbols you will be working with, it would be a good idea to learn to work with all types. However, if you know that you will only be involved with one type, then you should redraw the circuits, when necessary, using that type. It will be good practice in drawing circuits and working with symbols.

GLITCHES

Glitches are undesirable spikes that are on a desired signal. In logic circuitry glitches can result from the propagation delay in a circuit. By propagation delay we mean the time that it takes a signal to pass through a circuit.

The illustration in *Figure 1-9a* shows the input and output signals for an inverter. It is one of the logic gates that you will study later. The output signal is 180° out of phase with the input signal. At least, that is the theory of operation. In practice, it takes time for the signal to pass through the gate, so the phase difference is not exactly 180°. The delay may be only 20 nanoseconds, but that is enough to produce glitches in a later circuit.

As an example, suppose two out-of-phase signals are delivered to a circuit that has no output unless both inputs are positive; that is, at a logic 1 level. This is shown in *Figure 1-9b*.

If the signals are really 180° out of phase there will never be an output. But, if one of the input signals is obtained from an inverter there is a very short instant of time when both signals are positive. That instant of time is called the propagation delay. In this case it will result in undesirable spikes, or glitches.

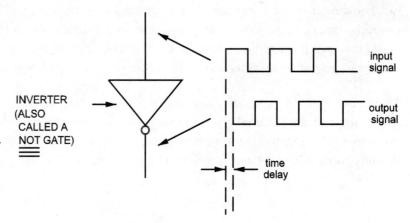

Figure 1-9a. Example of propagation delay is shown in this NOT inverter gate. The time delay is caused by the propagation delay.

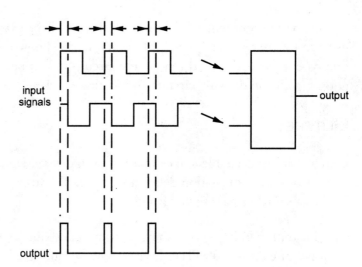

Figure 1-9b. The lower input signal has been delayed by a NOT gate. The output shows glitches.

IDEAS YOU *SHOULD* CARRY OVER FROM YOUR WORK IN ELECTRONICS

It is assumed that you are a beginning technician in some field. Therefore, you already have some knowledge that can be carried over to your work in logic circuits.

You probably know, for example, how regulated power supplies work. They are used extensively in digital circuitry because certain types of logic integrated circuits must operate from a regulated supply. Furthermore, all electronic logic systems require some form of DC supply.

The methods of using signal injection and signal tracing in analog systems will be reviewed. You will find that the same ideas are used in logic systems, but the types of test equipment used are not the same. As a technician, you are going to study the need for a systematic approach to troubleshooting. It is important to understand basic circuits in order to be able to find problems within a system.

With a little experience, you know how to find your way around a circuit board by looking for *landmarks*. For example, to locate C_{21} of *Figure 1-10* you would go to the center tap of R_{28}. That variable resistor may be a volume control or tone control of a radio, so it is easily located. In logic circuitry you will quickly learn to identify landmarks. Examples are ICs, LEDs, power supply leads and timing circuits.

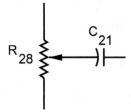

Figure 1-10. Use landmarks for locating components.

As with all electronic circuits, pins of devices such as ICs are numbered counterclockwise as viewed from the top. There is always an identifier to tell you the location of pin #1. Notches or dots are often used with DIP packages, and tabs are used on TO packages. DIPs are the Dual In-line Packages like the one in *Figure 1-11*. TO (Transistor Outline) packages are round metal packages used for transistors.

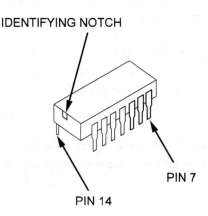

Figure 1-11. The location of pin 7 is shown for this DIP package. Pin 14 is also shown. Pins of electronic devices are numbered clockwise as when viewed from the bottom.

The Basis for Testing and Troubleshooting

Ideas You Should *Not* Carry Over From Your Work in Electronics

If you try to carry the following ideas over from your work in electronics you are going to make your job many times more difficult than it should be.

Don't try to analyze logic circuits by chasing electrons, or holes, around the networks. Many technicians are married to the idea that the only way to get insight into electronics is to find out where the current comes from and where it goes. In logic circuits it is just about impossible. There are integrated circuits that have over 10,000 transistors! There is one important exception. If you have a DC problem in the logic system you may want to use your knowledge of electron flow to trace DC paths and find the trouble.

You cannot troubleshoot digital circuits effectively with a voltmeter. Again, we make an exception in the case of trouble in the DC power supply circuit.

In a logic circuit you are usually interested in three basic signals:

> A higher-level voltage signal that is usually called logic 1. That doesn't mean 1 volt; it is just a way to refer to the higher voltage level.

> A lower-level voltage signal that is usually called logic 0. Again, it is not necessarily 0 volts; it is the lower voltage level.

> A pulsing voltage signal that alternates between 0 and 1.

The quickest and most convenient way to measure those signals is to use a logic probe, not a voltmeter. *Figure 1-12a* shows a drawing of a basic probe. The schematic drawing for the circuit of a more elaborate probe is given in Appendix A.

When the probe is touching a 1 signal, the LED is ON, and when it touches a 0 signal, the LED is OFF. When it touches a pulsing signal, the LED is on at half-brightness, or it is flashing ON and OFF. *Figure 1-12b* shows the symbols we use in this book for the different logic probe indications.

There are several reasons why a logic probe is better than a voltmeter in digital troubleshooting. You don't have to look up to take a reading each time you set the probe, and you don't have to interpret the readings you get. For example, suppose the voltmeter is used in a circuit where 4.5V is a logic 1 signal and 1 volt is a logic 0 signal. If a voltmeter indicates

2.5V, does it mean you are looking at a low 1 signal, or a high 0 signal, or a pulse? With a good logic probe that guesswork would be eliminated. The basic probe of Appendix A can usually tell the difference between a low value of logic 1 and a high value of logic 0. The probes that are sold commercially have internal logic circuitry that can make such readings more reliable.

Gain is no longer a factor in making logic circuit measurements. If there are two logic 1 signal levels into a logic gate, and a logic 1 signal level at the output, all of the 1's will be the same voltage value. An example is shown in *Figure 1-13*.

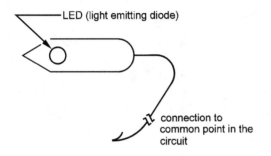

Figure 1-12a. A basic logic probe.

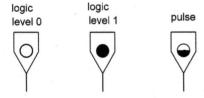

Figure 1-12b. Logic probe levels can be shown this way.

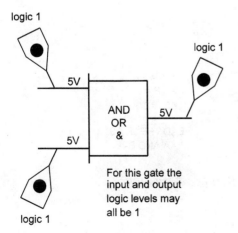

Figure 1-13. The two input levels and the output level are all at a digital level 1 in this case.

The Basis for Testing and Troubleshooting 19

THE BASIC GATES

There are six basic components that you will be working with: AND, OR, NOT, NAND, NOR, and EXCLUSIVE OR. When they are in integrated circuit form they are usually called *gates*.

There are four important features of these basic components that you must know: the schematic drawing symbol, the truth table, the circuit example and the math symbol, see *Figure 1-14*. It would be a good idea to memorize these features.

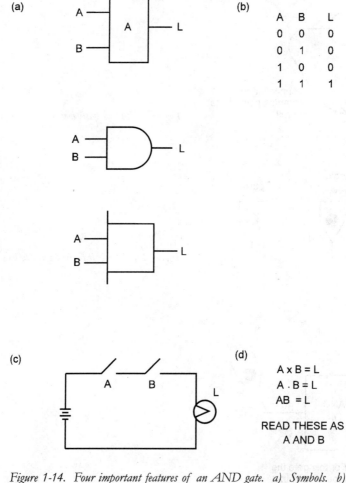

Figure 1-14. *Four important features of an AND gate. a) Symbols. b) Truth table. c) Circuit example. d) Math symbols.*

THE AND GATE

Three schematic symbols are given. The middle symbol is used most often in consumer electronic, military and industrial schematics. The lower symbol may be used on industrial diagrams. It would be a good idea to learn all symbols used for all the gates.

There are only two inputs (A and B) and one output (L) so it is called a two-input AND. It is possible for AND gates to have more inputs, but there is only one output terminal and one way to get a logic 1 output. That is, all inputs must be at a logic 1 level to get a logic 1 output. However, more than one connection can usually be made to an output. The number of inputs is called the *fan in* and the number of outputs is called the *fan out*.

The truth table lists all of the possible things that can happen at the input and output terminals. There are only two possible levels of voltage for each terminal, and they are represented by 0 and 1. Don't always think of 0 and 1 as being numbers. They can represent anything. A few examples are shown in *Table 1-2*.

For the AND gate truth table there are only two possibilities for each terminal: 0 and 1. The first row shows that if both inputs (A and B) are at 0, the output is at 0. The second and third rows show that if one input is 0 and the other is 1, the output is 0. The last row shows that if both inputs are at 1, the output is 1. It is important to note that the last column (L) shows that there is a 0 in the output at all times except when both inputs are at 1.

Motor	Not Running = 0	Running = 1
Lamp	Off = 0	On = 1
Tape Recorder	Not Recording = 0	Recording = 1
TV Set	Off = 0	On = 1
Relay	Not Energized = 0	Energized = 1
Transistor	Cutoff = 0	Saturated = 1
Voltage	0V = 0	5V = 1
Door	Open = 0	Closed = 1

Table 1-2.

The example circuit shows two switches in series with a lamp and a battery. The switches can be called the inputs, and the lamp the output. If you let 0 = Switch OFF, and 1 = Switch ON, then all of the possible combinations for OFF and ON are shown in the first two columns of the truth table. If you let 0 = Lamp OFF and 1 = Lamp ON, then the third column of the truth table shows that the lamp is always OFF unless both switches are closed. Every possible condition of the switches and the lamp are shown in the truth table. The trick is to represent the conditions of the switches and lamp with 0's and 1's.

THE OR GATE

Having gone into detail with the AND gate of *Figure 1-14*, it will be easier to discuss the remaining basic gates. The four important features of the OR gate are shown in *Figure 1-15*.

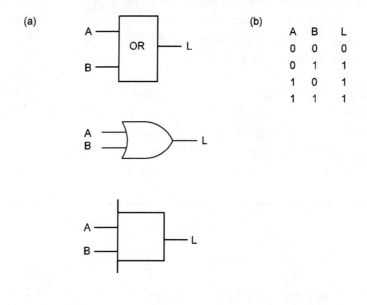

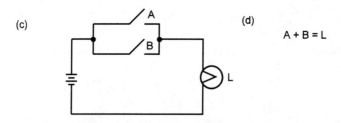

Figure 1-15. Four important features of an OR gate. a) Symbols. b) Truth table. c) Circuit example. d) Math symbol.

It would be a good idea to practice drawing the AND and OR symbols so that you will get them fixed in your mind.

The truth table and equivalent circuit show that there is an output (L) whenever there is a logic 1 at either or both inputs. That is sometimes called inclusive OR because it includes the possibility of both inputs being present. The following example will make this more clear.

Suppose an instruction sheet says: "The signal amplitude can be increased by adjusting R_3 OR R_4." Does that mean you can adjust both R_3 and R_4 to get more amplitude, or, does it mean that you can adjust either R_3 or R_4 but not both? If it means you can adjust R_3 or R_4 or both, then it is an inclusive OR situation. If it means to adjust one or the other, but not both, then it is an exclusive OR situation.

Note that the lamp will be ON in the circuit of *Figure 1-15* if either A or B is in the 1 condition, or, if they are both in the 1 condition.

Read A + B as A OR B in digital systems.

THE NOT GATE

Figure 1-16 shows the four important features of the NOT gate. With this type of gate the output is always the opposite of the input. So, if the input is 1 the output is 0; and, if the input is 0 the output is 1. The truth table shows the two possibilities.

In the relay circuit of *Figure 1-16* the lamp is ON whenever the circuit is NOT energized. That is, whenever switch A is open. When switch A is closed, the relay is energized and the lamp is OFF.

Any time there is a bar over a letter, such as $\overline{A}$, it means that the expression is negated. In other words, $\overline{AB}$ means NOT A AND B. An NOT gate is often called an inverter.

THE NAND GATE

The NAND gate is a Negated AND gate. You can see from the symbols in *Figure 1-17* that an NOT and an AND can be combined to get an NAND.

The truth table is obtained by inverting every term in the L column of an AND truth table.

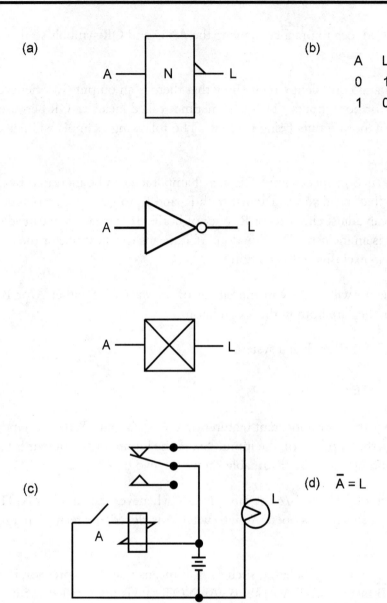

Figure 1-16. Four important features of the NOT gate. a) Symbols. b) Truth table. c) Circuit example. d) Math symbol.

That is very important because it shows that the output of two gates in combination can be obtained from their truth tables. The third column shows the output of the AND gate (AB) and the fourth column shows the output after the AND is inverted. The math symbol should be read "NOT A AND B."

24   *Electronic Troubleshooting and Servicing Techniques*

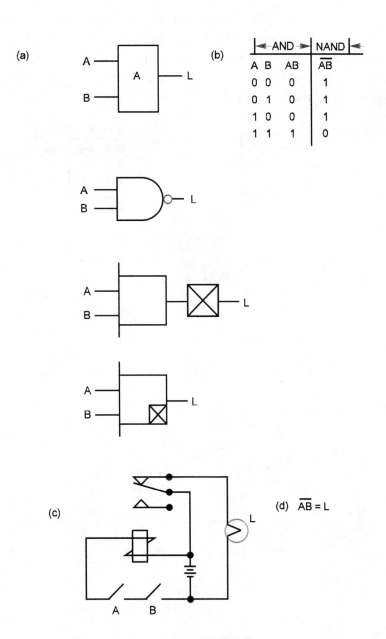

Figure 1-17. Four important features of the NAND gate. a) Symbols. b) Truth table. c) Circuit example. d) Math symbol.

The NOR Gate

As shown by the symbols in *Figure 1-18*, the NOR gate can be obtained by inverting the output of an OR gate. As with the NAND, the NOR output can be obtained by a truth table. Write the OR truth table first, then invert the output. *Figure 1-18c* shows another way to look at the NOR gate. Note that the switches are in an OR layout, and the NOT relay circuit is at the output of the switches. Read the symbol as "NOT A OR B."

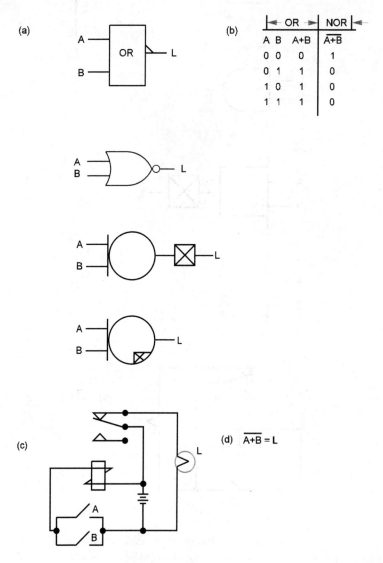

Figure 1-18. *Four important features of the NOR gate. a) Symbols. b) Truth table. c) Circuit example. d) Math symbol.*

The NAND and NOR blocks in *Figure 1-17* and *Figure 1-18* are very important. Manufacturers of integrated circuits often make NAND or NOR gates the basis of their complete line of gates. Because of their importance, you should spend some extra time studying these figures.

A two-input NAND gate can be used as a NOT. All you have to do is connect the two inputs together. Likewise, a two-input NOR gate can be used as a NOT. Again, all you have to do is connect the two inputs together, see *Figure 1-19*.

When you are troubleshooting logic circuits, you will run across connections like the ones shown in *Figure 1-19*. You should treat them as NOT gates. With a little experience you can interpret those circuits quickly.

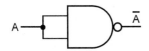

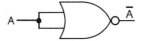

Figure 1-19. *By connecting the input leads together you can make NAND and NOR gates into NOT gates.*

THE EXCLUSIVE OR GATE

The exclusive OR features are shown in *Figure 1-20*. That type of gate means that for an output of 1 you can have either input in the 1 condition, but not both. Note that the output is zero whenever the two inputs are matched.

LOGIC FAMILIES

Before you have done much work in logic systems you run across the names of the various logic families. Examples are TTL, ECL, CMOS and (now obsolete) RTL. The names are actually descriptions of how the gates are made. For example, CMOS means that Complementary MOSFETs are used to make gates, and ECL means that the gates use Emitter-Coupled amplifiers as Logic gates.

It isn't necessary, or desirable, to spend a lot of time learning the electronic circuitry used to make the gates in each family. A two-input NAND gate does the same thing regardless of which family it comes from.

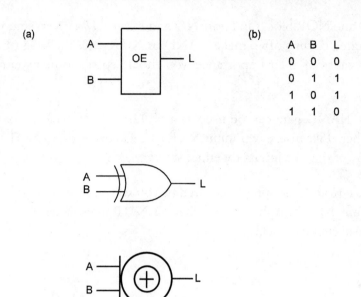
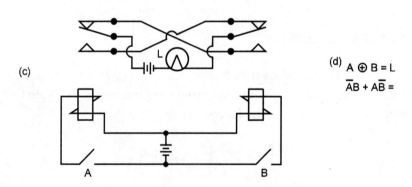

Figure 1-20. Four important features of the exclusive OR gate. a) Symbols. b) Truth table. c) Circuit example. d) Math symbol.

One reason for having the different families is that different manufacturers have their own preferred (patented) way to make gates. As a technician, there are certain identifying characteristics of families that you should know. They are listed below.

28 *Electronic Troubleshooting and Servicing Techniques*

TTL (Transistor-Transistor Logic)

This integrated circuit (IC) family of gates is very popular. A regulated 5-volt positive voltage power supply is *required* for operating TTL circuits. The +5V supply usually, but not always, goes to pin 14, and the common lead for the supply usually, but not always, goes to pin 7. The identifying numbers for the ICs with the basic gates usually start with 54 or 74. Examples: 5400 and 7400. The different numbers refer to identical gates, but one is specifically designed for military use.

ECL (Emitter-Coupled Logic)

ECL gets its name from the circuits used in its fabrication. In the ECL logic family you will find the same gates as used for the TTL and CMOS families. A distinguishing feature of ECL is its short propagation delay, which means it is the fastest of the four basic gates. So you cannot replace it with another logic gate from a different family. Another distinguishing feature of ECL is that it must be supplied with a negative power supply voltage. That is another reason ECL gates are not interchangeable with the gates of other logic families.

CMOS (Complementary Metal Oxide Semiconductor)

CMOS circuits use MOSFETs in their construction. Therefore, static charges can destroy older versions. Most of the newer CMOS gates are buffered. In other words, they have internal zener diodes at their terminals, which prevents destruction by static charges. The older gates, however, are not protected, so be careful with them.

CMOS logic ICs can be operated with any voltage from +5 to +12 volts, and the supply doesn't have to be regulated. That is a very important advantage over TTL. Their disadvantages are the possible destruction by static charges, and the fact that they are often slower in operation. The CMOS ICs often have numbers in the 4000's. Pins 7 and 14 are often used as power supply connections for 14-pin DIPs. For 16-pin DIPs, pins 8 and 16 are often used for the supply connections.

You may hear that a CMOS IC can be substituted for a TTL IC because the CMOS will operate on +5 volts. Don't you believe it! Even if the pinouts (pin numbering) are identical, they can't be interchanged because of the difference in their propagation delay. This is a measure of how long it takes for a signal to go through the gate; and, it is usually measured in nanoseconds. It takes longer to get a signal through a CMOS gate when it is operating at 5 volts. If it is interconnected with TTL gates, the signals just won't show up on time at some point down the line. To repeat, CMOS gates won't work in a TTL circuit!

A special series of CMOS chips have exactly the same pinouts as the equivalent TTL. They have a C in the middle of their number. As an example, a 74C76 is a CMOS integrated circuit that has the same pinout as a TTL 7476. Both are dual flip-flops, but they are NOT interchangeable.

RTL (Resistor-Transistor Logic)

RTL gates are not used in new circuit designs, but there are many still in use. They usually have identifying numbers in the 700, 800, and 900 range. Pin 11 is usually the positive supply connection, and pin 4 is used for ground. RTL ICs operate from a +3.6V regulated supply. The RTL ICs are usually made with gold-colored leads.

Manufacturers have booklets giving the pinouts for their logic families. CMOS is a Harris specialty, TTL is a Texas Instruments specialty. Other companies, like Motorola, make them under their own names (like McMos) and numbering systems. Unlike tubes and transistors where the identifying numbers, like 6L6 and 2N2905, are the same device regardless of which company makes them, logic ICs do not always have identical numbers. You wouldn't want to do much work in tube and transistor circuits without manuals that show the pinouts. It is even more important to get them for logic ICs.

TTL and CMOS are the logic circuits that you will most often encounter. For those and other types, the manufacturer of a system may or may not include the pinout and power supply information you need on their schematics. Remember that manufacturers may put their own identifying numbers on ICs. Also, remember that not all gates are in integrated circuit form. Some companies make their logic gates with discrete components (that is, from transistors, resistors, capacitors, etc.).

Summary

Troubleshooting starts with a visual inspection followed by analyzing symptoms and looking for obvious defects. Reports by the user may have limited value.

After the preliminary steps, the first measurement by test equipment should be the power supply voltage. If that isn't the correct value, don't go any further until it is fixed.

Technicians often use a statistical approach to locate a defective component. Other things being equal, the component that is easiest to replace is likely to be first on the replacement list. Components that are most likely to fail should be at the top of the replacement list unless your troubleshooting procedure indicates that some other component is at fault.

In the example given for power supply troubleshooting, the various methods of making measurements may be the key to locating the defective component.

There are two kinds of digital circuit measurements (qualitative and quantitative) with some overlapping of the two. It is important to understand the type of measurement you are making. Qualitative and quantitative methods have been discussed in the analog section of this chapter.

Technicians must know the capabilities and the limitations of their test equipment. Use of the wrong instrument can lead to a wrong diagnosis. Although it is very important to know how to use a wide variety of test instruments, the most efficient approach is to use the least amount of equipment necessary to get the job done properly.

Chapter 1 Quiz

1. When a measurement provides a number value, such as 3V or 27K, it is an example of a _____ measurement.

2. When a measurement does not provide a number value, but instead provides information for making a decision, it is an example of a _____ measurement.

3. A voltmeter is connected across an On/Off switch. It reads zero volts when the switch is open. Does that mean the switch is OK?

4. The power amplifier transistor in a certain system is shorted from the emitter to the collector. Would that make the power supply voltage higher or lower than normal?

5. A current meter is connected in series with an N-channel JFET circuit being measured. Should the positive side or negative side of the meter go toward the drain?

6. Name two kinds of voltages that can destroy an electrolytic capacitor.

7. An ohmmeter is being used to determine if an electrolytic capacitor is passing a leakage current. Is this a proper test?

8. A fuse can be checked with a voltmeter. Connect the meter

   A. in parallel with the fuse.

   B. in series with the fuse.

9. After preliminary checks, the first troubleshooting measurement with test equipment should be to measure the _____ voltage.

10. A certain defective rectifier diode conducts equally in both directions. In a power supply that could result in the destruction of _____.

11. The surge-limiting resistor (R_1) in a certain half-wave rectifier (like the one in *Figure 1-1*) has burned out. Which of the following is correct?

   A. That could be the result of a defective electrolytic capacitor.

   B. That problem cannot be caused by a defective electrolytic capacitor.

12. Can a defective diode destroy the electrolytic filter capacitors?

   A. Yes

   B. No

13. Refer to *Figure 1-1*. Is the following statement correct? You can determine the condition of an electrolytic capacitor by measuring the DC voltage across it with an analog voltmeter.

    A. The statement is correct.

    B. The statement is not correct.

14. You have accidentally installed the rectifier diode backwards in the circuit of *Figure 1-1*. Which of the following is correct?

    A. Parts of the circuit will be destroyed.

    B. The diode will be quickly destroyed. So no other damage will result.

15. A square wave test is being conducted on an amplifier. It is an example of a

    A. qualitative test.

    B. quantitative test.

    C. test that can be either qualitative or quantitative.

    D. test that is neither qualitative nor quantitative.

16. After looking for obvious defects and considering the symptoms, the first actual measurement with test equipment should be to

    A. check the condition of the ON/OFF switch.

    B. measure the power supply voltage.

17. The output voltage is measured across the load resistor in the circuit of *Figure 1-1*. It is found to be only one-fourth the required output voltage. Which of the following would be a logical step to follow?

   A. Check the resistance of the filter resistor.

   B. Disconnect the output line of the power supply.

18. If the AC input voltage to a half-wave rectifier circuit (like the one in Figure 1-1) is 100V, what output voltage could you expect to measure across R_L?

   A. 92V

   B. 141.4V

19. An analog VOM is connected across a 6.3V winding on a filament transformer. The meter is *incorrectly* set to the 50V DC scale. Which of the following is correct?

   A. The meter will display zero volts.

   B. The meter will be destroyed.

20. As a general rule, should you stand an analog meter upright if it has a jeweled movement or lay it on its back when measuring voltages?

21. Consider a battery with its internal resistance. Can you measure that resistance by connecting an ohmmeter across terminals?

22. Which type of meter can be used to give an accurate measurement of an electrolytic capacitor's condition?

23. Write the equation for the rise time of a square wave on an oscilloscope used to determine bandwidth.

24. You want to measure the voltage across a 15 megohm resistor. You should use a(n) _____ voltmeter or _____.

25. An oscilloscope with a narrow bandwidth of 20 MHz rating is liable to miss a

    A. glitch.

    B. logic level.

26. What are the only logic levels used in digital logic and microprocessor circuits?

    A. Forward and reverse (also called Ahead and Back)

    B. 1 and 0 (also called High and Low)

27. What is the output of the circuit in *Figure 1-21*?

    A. A or B and C

    B. A and B or C

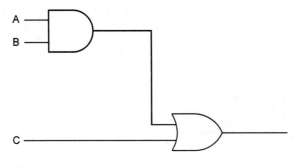

Figure 1-21.

28. Which logic family will NOT work with a +5 volt supply?

    A. CMOS

    B. ECL

    C. TTL

29. When there is a bar over the letter M it means _____ _____.

30. Name three of the most popular integrated circuit families.

31. Which gate is used to make an enable?

32. Identify the gate that has the following truth table.

    0 0 1

    0 1 0

    1 0 0

    1 1 0

33. A certain radio can only tune in one station. After a good visual inspection shows no obvious problem, your first measurement should be _____.

34. The term *power supply load* means

    A. the power supply output resistance.

    B. current delivered by the power supply.

CHAPTER TWO

TROUBLESHOOTING BASIC ANALOG CIRCUITS

OVERVIEW

A very important part of troubleshooting is making measurements. The usual procedure is to make a measurement, compare the result with what you should get, and then make a decision as to whether or not a problem exists at the point you are measuring.

That procedure cannot work unless you know exactly what measurement you are supposed to get. In some cases, the necessary information is given on the manufacturer's schematic, or it is given in the technical literature for the particular product you are servicing.

However, there is much information you are expected to know, and you cannot afford to look up that information each time you make a measurement. Some good examples are the polarity of the voltages on electrodes of the amplifying devices, signal configurations for those amplifying devices, and the methods for connecting devices to a DC source and the inputs and outputs of logic gates. Those are some of the subjects reviewed in this chapter.

This chapter is not intended to be a replacement for technical training in electronics. It is more of a review, that can be used as a supplement to basic training in electronics. Only those basic concepts that are directly related to troubleshooting are covered. Linear devices and basic digital devices are both discussed in this chapter. Basic operational amplifier (op amp) material is included in later chapters.

A typical method of teaching about the amplifying devices used in linear circuits is to show the differences between each device. That concept is not used here. Instead, we review the similarities of the devices, making it easier to group them into categories.

Some of the linear topics discussed include:

> What are the polarities of voltages to be expected on the various linear amplifying devices?
>
> What is the importance of amplifier configuration in troubleshooting?
>
> How are DC connections made to amplifying devices?
>
> How are the methods of coupling related to the types of amplifiers you are working with?
>
> What are the important methods of biasing linear devices?
>
> What are the important characteristics of power amplifiers that simplify their servicing?
>
> How does a superheterodyne radio serve as an example of systems used in linear troubleshooting problems?
>
> How do you test operational amplifiers?
>
> How do you use an ohmmeter to test a transistor?

DC Voltages on Amplifying Devices

Figure 2-1 shows the basic amplifying devices, their related electrode terminology and polarities of voltages to be expected on each device. All are three-terminal devices. That means that are all similar with respect to their control terminal, and input and output terminals. All can be represented by the rectangular boxes shown below the schematic symbols.

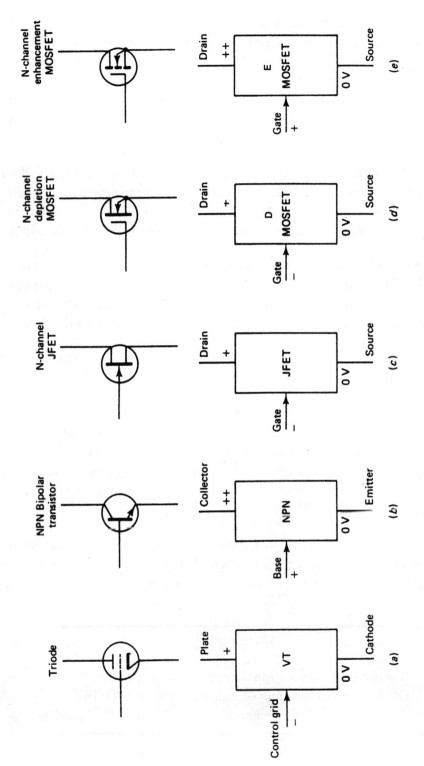

Figure 2-1. Basic amplifying devices showing their electrode terminology and the voltage expected on each electrode.

Note that the control electrodes are called control grid, base or gate. Note also that all of the voltage polarities are taken with respect to the cathode, emitter or source. For example, in a bipolar transistor the voltages would be taken with respect to the emitter, and in a vacuum tube the voltages would be taken with respect to the cathode. For a field effect transistor, the source is considered to be the reference voltage with respect to the other electrodes. Those voltages are considered to be zero volts with respect to other electrodes regardless of their DC polarity in relation to the chassis or other components.

The NPN and PNP transistors are bipolar devices. They are different from all the other amplifying devices because they require a base current in order to get a collector current. All of the other devices require a voltage on their control electrodes. Remember: the arrows on all semiconductor symbols point away from a P region and toward an N region.

What actually happens inside the device is of no consideration when you are troubleshooting. If the amplifying devices are used in a conventional way you can expect the input signal to be on the control electrode and the output signal to be either at the DC input electrode (cathode, emitter or source) or the DC output electrode (plate collector or drain).

In the semiconductor devices there are alternate components made by interchanging N- and P-type materials. That cannot be done in a vacuum tube. *Figure 2-2* shows the alternate devices and the polarities of voltages to be expected.

Returning to *Figure 2-1*, note the similarity in the polarities of voltages between the bipolar (NPN) and the N-channel enhancement MOSFET. Since their polarities are similar, you can expect similar methods of biasing to be used. In the same way, the vacuum tube, JFET, and N-Channel MOSFET also have similar polarities, and their methods of bias are very much the same.

The enhancement MOSFET must have a forward voltage bias before a drain current can be obtained and you should use some caution when working around those devices. Like the vacuum tubes, they may be operated at very high DC power supply voltages.

Warning

It is not uncommon to find a voltage of 400 or 500V DC supplied to the drain of enhancement MOSFETs. The same is true of vacuum tube plates.

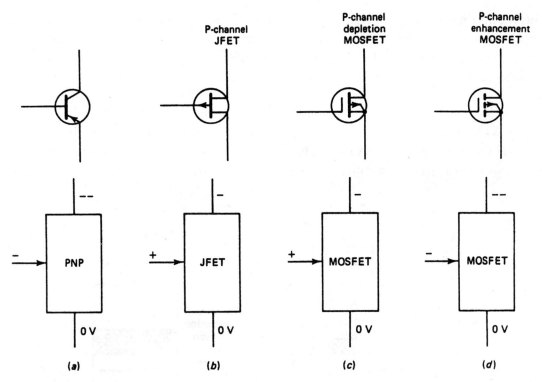

Figure 2-2. These are the devices that have no vacuum tube counterpart. Note the difference in symbols, especially the directions of the arrows.

Technicians sometimes get complacent when working around solid-state equipment because those devices are normally operated at low voltages. The enhancement MOSFET is a solid-state device and very often is an exception to that rule.

There are also some bipolar transistor amplifiers that are operated with a high voltage (about 100V) on the collector. They are used in high-frequency circuits, and the reasoning can be understood by referring to *Figure 2-3*.

Recall two important basic theories about bipolar transistors. First, the base and collector junctions are reverse biased in the normal operation of a bipolar transistor in an amplifier circuit. Although the base and collector voltages of an NPN transistor are both positive, the collector is more positive than the base. Therefore, the base-to-collector junction is reverse biased. Likewise the base and collector junction of a PNP transistor are both negative, but, since the collector is more negative than the base the base-to-collector junction is reverse biased.

Troubleshooting Basic Analog Circuits 41

This means there is a depletion region between those two sections. The depletion region is a no-man's land where very few charge carriers exist. That junction can be thought of as semiconductor region. So the junction behaves like a capacitor because it has N and P conducting materials separated by an equivalent dielectric.

Recall, also, that the capacitance of a capacitor varies inversely as the distance between the plates. With a very small distance (or, a thin depletion region), the capacitance between the base and collector can be quite high. A high capacitance permits high-frequency signals to sneak through the transistor without being amplified. The path is shown by the arrow in *Figure 2-3a*.

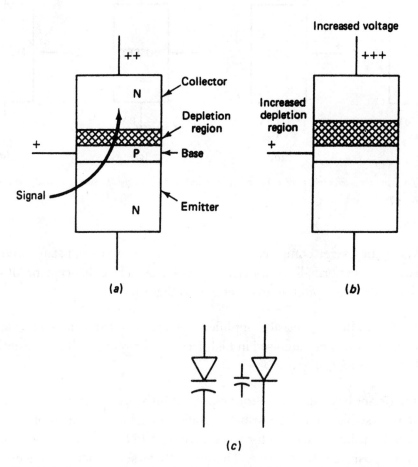

Figure 2-3. Junction capacitance in a bipolar transistor. a) The arrow shows how the signal sneaks through the transistor capacity without being amplified. b) Decreasing the capacity by increasing the reverse voltage between the base and collector. c) Symbols for varactor diodes. These diodes use the depletion region as an insulator.

If you increase the reverse bias between the base and the collector, the depletion region is increased. That is shown in *Figure 2-3b*. Increasing the depletion region is the same as moving the plates of the base-to-collector capacitance further apart. In other words, the capacitance is decreased.

In an RF circuit it is quite possible for the collector of the bipolar transistor to be operated at a high positive voltage in order to reduce the junction capacitance. The transistor must be designed so that it is capable of withstanding the increased reverse bias.

This discussion of the junction capacitance also applies to the varactor diode. Its symbol is shown in *Figure 2-3c*. In that device the junction diode is reverse biased and it is operated as a capacitor rather than as a diode. Varactor diodes are used extensively in tuning circuits and voltage-controlled oscillators.

Figure 2-4 shows some polarities of voltages on devices that you will not see very often in your troubleshooting work. The input signal on the tetrode and pentode is at the control grid and the output signal is at the plate.

When the dual-gate MOSFET is used as an RF amplifier one gate is used for the signal and one gate is used for AGC. When used as a converter, one gate is for the oscillator signal and one is used for the RF signal. The DC gate voltages depends upon the application so the polarity and magnitude can vary. However, you should be able to make the proper DC measurements when they are encountered.

Go back again and review the symbol for the triode tube in *Figure 2-1*. The control electrode (or control grid) regulates the number of electrons that go from the cathode to the plate. The tube amplifies because a very small change in control grid voltage causes a relatively large change in the cathode-to-plate current.

Again, note that the control grid and plate are two metal objects separated by a vacuum and, therefore, they form a capacitor. That capacitor, as in the case of the bipolar transistor, permits the signals to sneak through from the grid to the plate without being amplified.

To eliminate the effect of the grid-to-plate capacitance, an additional grid is added to the tube, as shown in *Figure 2-4a*. The screen grid shields the control grid from the plate. It acts like a Faraday screen, or a Faraday shield.

The screen grid is operated with a positive DC voltage. That helps to draw charge carriers (electrons) away from the cathode, which, in turn, increases the plate current.

There is a problem with the tetrode of *Figure 2-4a*. Under certain conditions, high-velocity electrons striking the plate bounce off and go to the positive screen grid. That, in turn, results in a decrease of plate current.

Since an increase in plate voltage causes a decrease in plate current (due to the increase in secondary electrons), the tetrode is said to have a negative resistance over a range of plate voltages.

To eliminate that condition, a third grid, called the suppressor grid, is added between the screen grid and the plate. The result is shown in the pentode of *Figure 2-4b*. Usually, the suppressor grid is connected internally to the cathode in the tube. That makes it negative with respect to the plate. The negative voltage on the suppressor grid drives secondary electrons back to the plate and prevents the reduction in plate current over certain parts of the characteristic curve. If the suppressor lead is brought outside the tube for external use, it has a negative bias voltage.

Although vacuum tubes have yielded their first-place position in electronics there are still a few applications for the tubes. They still have applications in high-power transmitters and some high-priced audio systems.

The second gate in the dual-gate MOSFET, the one furthest from the source, is often used for an AVC or AGC input. The first and second gates are also used for mixing two signals.

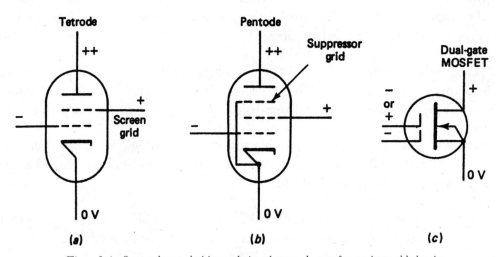

Figure 2-4. *Some voltage polarities on devices that you do not often see in troubleshooting.*

When you are making DC measurements around a dual-gate MOSFET, remember that the second gate can be positive or negative under different conditions. It may be necessary to consult the schematic or investigate the circuit further to determine if the polarity on that electrode is correct.

All of the devices in *Figure 2-1*, *Figure 2-2* and *Figure 2-4* are shown in another form in *Figure 2-5*. The amplifying devices are represented by models. Having a clear view of models sometimes makes it easier to understand the operation of the devices. Let's briefly review the models.

Models

The symbol for the triode tube shown in *Figure 2-5a* also serves as a model for its operation. The cathode is heated until electrons are *boiled* off the surface. These electrons are attracted to the plate.

The control grid between the cathode and the plate is given a negative voltage with respect to the cathode. The more negative the voltage, the fewer the number of electrons reach the plate. Therefore, applying a negative-going signal to the control grid (which is not allowed to go positive) makes it possible to control the electron stream in accordance with the input signal.

The key to amplification by the triode is that *a small change in voltage on the grid produces a relatively large change in plate current*. When a screen grid and a suppressor grid are added to this amplifying device they do not affect the basic operation described here. So tetrodes and pentodes are still considered to be three-terminal active devices. (An active device is one that can be considered to be the source of a new signal.)

The NPN bipolar transistor model, illustrated in *Figure 2-5b*, is made by sandwiching a P-type material between two N-type materials. Note that the collector is more positive than the base in the NPN transistor. In other words, the base is negative with respect to the collector.

The charge carriers are electrons. Most go from the emitter to the collector, but some go to the positive base. Increasing the base current greatly increases the number of electrons that go from the emitter to the collector.

The key to amplification for this device is that *a small change in base current produces a relatively large change in collector current*.

Figure 2-5c shows the model for an N-channel JFET (Junction Field Effect Transistor). The large arrow represents the motion of charge carriers (electrons in this case) through the N-channel of the JFET. The gate, made with P-type material, is reverse biased with respect to the channel. Note the positive voltage on the drain and the negative voltage on the gate. With that reverse bias there is a depletion region around the gate and that limits the amount of area through which the charge carriers can pass. Increasing the negative gate voltage increases the depletion region and decreases the amount of current that can flow from source to drain. The depletion region is represented by the shaded area.

The key to the operation of the N-channel device is that *a small change in the gate voltage results in a relatively large change in drain current.* The substrate can be thought of as being the material upon which the JFET has been constructed. In a normal operation, the substrate is a semiconductor material with very little conductivity. It is usually operated at the source or ground potential.

Figure 2-5d shows the model for the N-Channel MOSFET. It is very much like the one for the JFET with one exception. There is a black region around the gate that represents an insulating material. Originally those devices were called IGFETs (Insulated Gate Field Effect Transistor). That name is still used in a few applications. The insulating material is made of a metal oxide, hence the name: Metal Oxide Semiconductor Field Effect Transistor (MOSFET).

As with the JFET, the negative voltage on the gate reverse biases the PN region and produces a depletion region. That limits the amount of current flowing from the source to the drain. The depletion region is represented by the shaded area. Making the gate more negative decreases the drain current because of the resulting increase in the size of the depletion region.

As with the JFET, *a small change in the gate voltage causes a relatively large change in the drain current.* The operation of the gate is such that it depletes the size of the conducting region.

At this point we need to fine-tune our discussion of the MOSFET action. We have discussed the MOSFET by using a convenient model of a depletion region. Actually, there is an electric field between the gate and the substrate. Hence the name *field effect*. That field is what actually controls the flow of electrons through the device.

However, the important thing to remember is that a negative voltage on the gate of an N-channel MOSFET controls electron flow through the channel. Use whatever model of behavior you prefer.

Figure 2-5e shows the model for an N-Channel enhancement MOSFET. The obvious difference between this device and the depletion MOSFET is that the depletion region reaches all the way through the channel. So, when there is no positive voltage on the gate, there can be no current from the source to the drain.

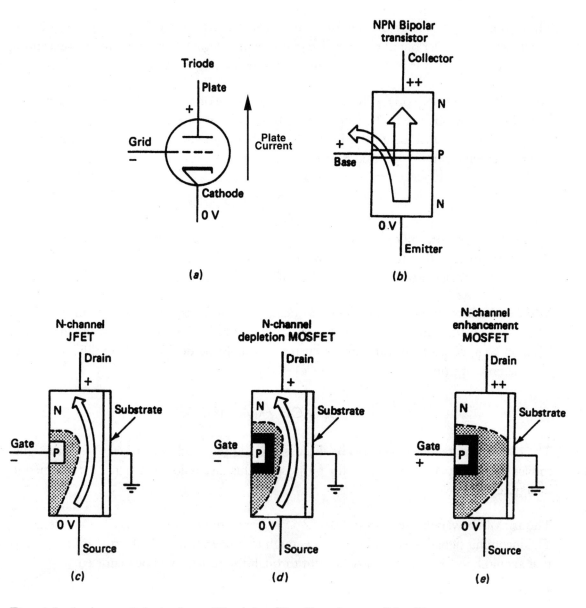

Figure 2-5. Another way of showing the amplifying devices. These illustrations are called models.

In order to start current flow from the source to the drain, it is necessary to forward bias the gate. That reduces the gate-to-drain voltage and thereby reduces the size of the depletion region. In turn, that permits current to flow through the channel.

Again, observe the similarity between the voltages on the electrodes of the NPN bipolar transistor and the N-channel Enhancement MOSFET. Note also the similarity and voltages on the electrodes of the vacuum tube, JFET and depletion MOSFET.

These models are not to be thought of as methods of constructing the device. In reality, the construction is similar but not exactly like that shown in *Figure 2-5d*. However, by presenting the models this way it is easy to visualize what is happening in the amplifying devices.

Power MOSFETs were originally called V-FETs because of the way they were constructed. V-FETs are now obsolete. They are usually shown with the same symbol and same polarities as used for enhancement MOSFETs. Remember, they are capable of delivering power and they are likely to be operated with high voltages.

A Resistor Model

As mentioned before, all of the amplifying devices discussed so far are three-terminal devices. That means that the signal is applied to one of the electrodes, usually the control electrode. That signal controls the number of charge carriers that pass *through* the device.

You can think of an amplifying device as being a variable resistor, as shown in *Figure 2-6*. The signal on the control electrode moves the arm of the variable resistor up and down. That, in turn, controls the amount of current through the device. All amplifying devices operate on that principle.

Classes of Amplifiers

The classes of operation will also be discussed using the models in *Figure 2-6*. The signal moves the arm up and down. For the purpose of this discussion, the signal is presumed to be a pure sine wave.

The arm of the variable resistor in *Figure 2-6a* is shown in the position for Class A operation. The incoming signal must not be strong enough to move the arm all the way to the end. If that should happen, the sine wave current through the resistor will be distorted.

In the Class B operation of *Figure 2-6b*, the resting point of the arm is at the bottom of the variable resistor. Now the applied sine wave can move the arm up, but it cannot move the arm down. Therefore, only the positive half cycles - during which the arm is moved up - will affect the current through the resistor. As shown by the current waveform, the current increases as the arm moves up on the positive half cycle but cannot decrease when the arm moves down.

The current waveform in Class B operation is not a perfect representation of the input sine wave. There is a considerable amount of distortion. However, note that the arm can now move throughout the total distance of the variable resistor on the positive half cycle. Therefore, a greater amount of control is exerted over the conducting half cycle.

Some of the amplifying devices can be operated Class C, as represented in *Figure 2-6c*. Note that bias voltage is some distance below zero volts, and that the quiescent point of the signal is also below zero volts. In order to get control of the output current, it is necessary for a certain amount of the signal to be used to get the arm up to the zero-volt point. Above the zero point, the arm can move throughout the variable resistor.

Class C operation is valuable in some circuits. For example, in a vacuum tube circuit, Class C operation is characteristic of oscillators and high-power amplifiers. It is not used in bipolar transistor circuits because in the quiescent condition the emitter-base junction would be reverse biased. As a rule, bipolar transistors are not designed for operation with a reverse bias on the emitter-base junction.

Under certain conditions, the JFETs and MOSFETs may be operated with Class C operation but that type of operation has not been used extensively.

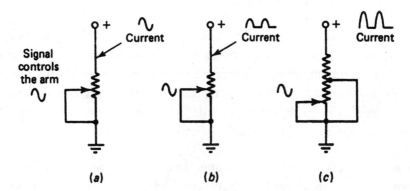

*Figure 2-6. a) In Class A operation the input signal increases and decreases the resistance.
b) In Class B operation the input signal only decreases the resistance. c) In Class C operation a certain amount of signal is needed to get the resistance up to zero ohms.*

When the three-terminal device is operated in a conventional manner, as indicated in *Figure 2-6*, the input signal *voltage* is to the control electrode and the output signal voltage at the DC output electrode is 180 degrees out of phase with that input signal voltage. The output signal *current* is in phase with the input signal voltage.

CONFIGURATIONS

When you remember that the amplifying device is a three-terminal device, it should be obvious that one of the terminals has to be held constant with respect to the other two. The input signal is taken from some point relative to common and the output signal is taken from some point relative to common. Therefore, one of the three electrodes must be common. That permits three *configurations* of operation as shown in *Figure 2-7*.

If the input signal is to the control electrode and the output is taken from the DC output terminal, then you have a common emitter, common source or grounded cathode configuration, see *Figure 2-7a*. It has the advantage that it has the best compromise between voltage gain and power gain. Also, it gives satisfactory operation at high frequencies.

In *Figure 2-7b*, the input signal is to the DC input electrode and the output is taken from the DC output electrode. The control electrode is common. Since the control electrode is at zero volts, it acts as a shield between the input and output signals. Therefore, there is no way for the signal to sneak through the device without being amplified. This kind of configuration is called grounded grid, common base, or common gate. It does not have the high-voltage gain available in the conventional operation of *Figure 2-7a*. The input impedance of the circuit is low and the output impedance is high, so it can be used as an impedance-matching device. Whenever you see it, you should always consider it is very possible that you are working in a high-frequency circuit.

Figure 2-7c shows a follower configuration. The input signal is on the control electrode, and the output is on the DC input electrode. This follower circuit has a high input impedance and a relatively low output impedance. There is a voltage gain of less than 1, but it does have a power gain. It is also possible that it is being used to match a high-impedance circuit to a low-impedance device or circuit. For example, it is sometimes used to match the high impedance of an amplifier to a low-impedance transmission line.

Follower circuits are used also for level shifting. In direct-coupled devices, the DC level of the signal (sometimes referred to as DC offset) gets higher and higher from device to device. The follower circuit brings the DC level out of the signal down to a point where it is close to common, or zero volts. Thus the name *level shifting*.

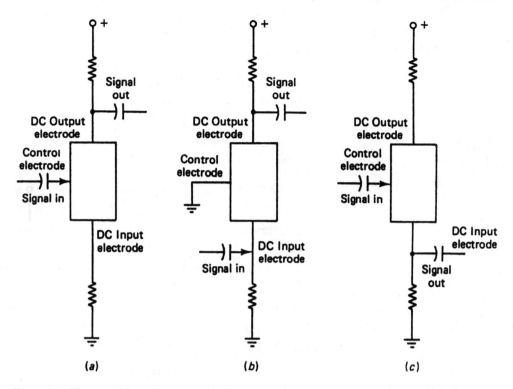

Figure 2-7. Three amplifier configurations. a) A conventional, or common emitter configuration (also known as common source or ground cathode). b) This configuration is called grounded grid, common base or common gate. It is most often used for high-frequency operation. It is also used to match a low impedance to a high impedance. c) The follower configuration. It is primarily for matching a high impedance to a low impedance.

When you are troubleshooting, you must understand what the configurations are used for and how they are affected by their DC connections. Note this very important point: These are signal input and output configurations. They may have nothing to do with the DC power connections to the device. Just as one of the terminals must be common for the input and output signals, it is also true that one of the electrodes must be common for the input and output DC voltages.

DC Supply Connections

In *Figure 2-8* the common symbol is used to represent zero volts. In the conventional connection of *Figure 2-8a*, the DC input electrode is grounded and the output electrode goes to the power supply voltage. The power supply is marked positive, so it is assumed that it is one of the devices shown in *Figure 2-1*. If it was one of the devices in *Figure 2-2*, the power supply connection would be negative with respect to common.

In *Figure 2-8b*, the DC output electrode is common and the power supply is delivered to the DC input electrode. Note the negative supply polarity.

In the third configuration, *Figure 2-8c*, the DC input and output electrodes are delivered to separate power supplies, one positive and other negative. That configuration is sometimes referred to as positive/negative bias or long-tail bias. Since the voltage makes the transition from negative to positive through the device, it stands to reason that the control electrode is either at, or very near to, ground potential.

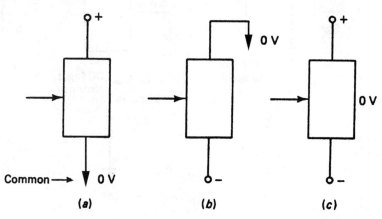

Figure 2-8. Various ways to connect the power supply to an amplifying device. a) The supply connected to the current output electrode. b) The supply connected to the current input electrode. c) Long-tail bias using a positive and negative supply.

The DC configuration of *Figure 2-8a* demonstrates the importance of always considering the DC input electrode to be zero volts. That way, the polarities of the voltages on the other electrodes will still be correct.

The configuration you will most often see is grounded (or common) cathode, common emitter and common source. It is important to remember the phase signals for this configuration. In *Figure 2-9*, note that when the input signal is at the control electrode, the output signal voltage is 180 degrees out of phase with the input signal voltage. The signal at the input electrode is in phase with the incoming signal.

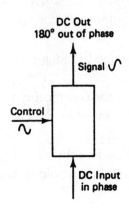

Figure 2-9. The 180° phase shift for a conventional amplifier.

Methods of Obtaining Bias

A DC bias voltage must be applied to an amplifying device in order to get it into the quiescent condition. *Figure 2-10* shows four kinds of bias.

AGC (or AVC) bias *Figure 2-10a* is used primarily for receiver amplifiers. The bias voltage is obtained by rectifying and filtering the incoming signal.

Battery bias in *Figure 2-10b* is used in portable equipment and in equipment where it is absolutely necessary to have no power supply hum and no interference on the control electrode.

In *Figure 2-10c*, a separate power supply is used for bias. That kind of bias is used extensively with high-priced, high-quality regulated power supplies. It is also used extensively in power amplifiers where a high bias voltage is needed.

When the DC polarity of the control electrode is opposite to the polarity of the DC input electrode, self bias can be used. An example using a JFET is shown in *Figure 2-10d*. The DC current through the amplifying device also flows through the input resistor (R_1). That makes the DC input electrode (the source in this case) positive with respect to common. Since there is no gate current, it is at zero volts. That is another way of saying the gate is negative with respect to the source. So, the N-channel JFET is properly biased.

Self bias, shown in *Figure 2-10d, cannot* be used with bipolar transistors and enhancement transistors because the DC voltage on their control electrode (base or gate) has the same polarity as the voltage on their base or source.

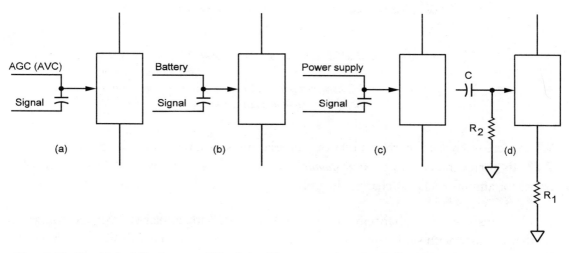

Figure 2-10. Four kinds of bias for an amplifying device. These types can be used with all amplifying devices discussed thus far.

AGC/AVC bias will be discussed in reference to the radio system described later in this chapter. However, there is one point that should be made at this time. A special case for AGC bias will be noted, and it is illustrated in *Figure 2-11*.

There is a decrease in gain in most amplifiers as the bias is moved toward cutoff. In an N-channel JFET, for example, making the bias more negative will decrease the gain over a range of bias values. That characteristic is shown in *Figure 2-11a*.

For the bipolar transistor, increasing the forward bias will increase the gain up to a point. After that, the gain decreases as the forward bias increases.

Suppose, for example, that the graph of *Figure 2-11b* is for an NPN bipolar transistor. Forward bias is accomplished by making the base positive with respect to the emitter. As the base voltage is made more and more positive, the gain increases up to a point. At some value of forward bias, the gain is maximum. After that, a further increase in forward bias results in a decrease in gain.

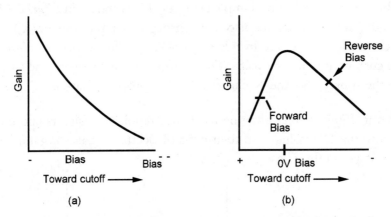

Figure 2-11. Two types of AGC bias. a) For conventional AGC bias, the gain increases as the bias decreases. In this illustration the gain is decreasing with an increase in bias. b) With bipolar transistors, there is a certain point of bias for maximum gain. Any change in bias from that point will result in a decrease in gain.

When there is a resistor in the emitter of a bipolar transistor, it is *not* used for bias, see *Figure 2-12*. Its correct name is *temperature stabilizing resistor* because it helps to stabilize the bipolar transistor amplifier against changes in ambient (surrounding) temperature.

Bipolar transistors and enhancement MOSFETs operate with similar DC voltage polarities. So they are used with similar polarities of bias. A popular way to get that bias is shown in *Figure 2-13*.

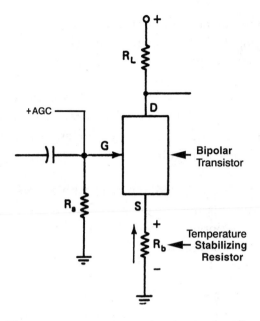

Figure 2-12. The setup for self-bias. This type of bias is used with tubes and field effect transistors.

In *Figure 2-13a*, a voltage divider is used between the positive power supply voltage and common. That voltage divider forward biases the base of the bipolar transistor (or, the gate of an enhancement MOSFET). Voltage divider bias is the type you will most often encounter for those devices.

There is a trade-off that is characteristic of all amplifiers and should be understood by a troubleshooting technician. That trade-off is between the amount of gain of an amplifier and the range of frequencies it can pass. In other words, amplifier bandwidth and gain are trade-offs.

Anything you do to increase the gain will automatically decrease the bandwidth. Conversely, *anything you do to decrease the gain will automatically increase the bandwidth.*

A good way to remember that is if you decrease the gain of the amplifier down to the point where the output and input signals are the same, then the amplifier has the same gain as a piece of straight wire, and it has a very wide bandwidth.

One way of decreasing the gain of a bipolar amplifier is to connect the base bias voltage divider circuit directly to the collector, as shown in *Figure 2-13b*. For a MOSFET it would be connected to the drain. In either case, the voltage divider is taken from a point where the output signal voltage is 180° out of phase with the input signal. Part of the output signal is fed back to the input electrode by the voltage divider. That out-of-phase signal partially cancels the input signal. The result is a decreased gain and a resulting increased bandwidth.

Another common way of decreasing the gain is illustrated in *Figure 2-13c*. A resistor at the DC input electrode is used. All of the amplifying devices can be operated with the resistor at the DC input position. The signal at that input is in phase with the incoming signal.

So, when the incoming signal goes positive, the voltage drop across the resistor increases, causing the DC input electrode to follow the signal. The DC electrode follows the signal up and down and decreases the signal voltage between the control electrode and the input voltage of the amplifier. That, in turn, increases its bandwidth.

If the switch in *Figure 2-13c* is closed, a capacitor is connected in parallel with the resistor. The capacitor charges and discharges and maintains the voltage across the resistor at a constant value. Now the input voltage at the DC input electrode is held constant. That increases the gain of the amplifier, but its bandwidth is decreased.

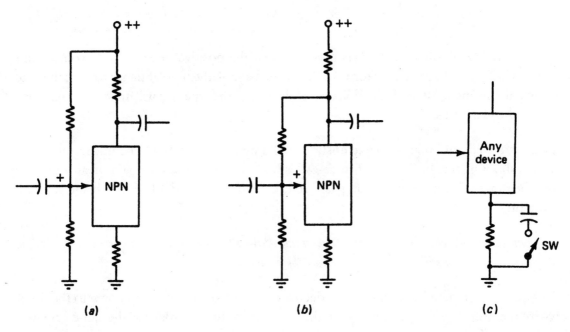

Figure 2-13. Types of bias for bipolar transistors and enhancement MOSFETs. a) A basic voltage divider bias. b) Voltage divider bias with negative feedback. c) Another way of getting negative feedback. In this case an open switch results in negative current feedback for the device.

A basic bias circuit for bipolar transistors is shown in *Figure 2-14*. It is relatively unstable with temperature changes compared to voltage divider bias. It is used in low-cost circuits that use bipolar transistor amplifiers.

The emitter resistor is used in bipolar transistor circuits for stabilizing against temperature changes. Without it, a thermal runaway condition can occur. Thermal runaway is an upward-spiraling condition in which an increase in a transistor's temperature causes its current to increase, and an increase in its current causes its temperature to increase. It rapidly destroys the transistor. *Remember, the emitter resistor is* not *a bias resistor.*

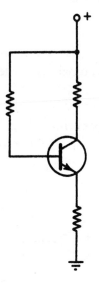

Figure 2-14. A basic bias circuit. This circuit is used in toys and very low-cost equipment. It is not the best way to bias a bipolar transistor.

COUPLED CIRCUITS

There are four methods of coupling the signal from one amplifier to the other. They are shown in *Figure 2-15*. Coupled circuits can be the clue to the kind of system you are troubleshooting.

The R-C coupling shown in *Figure 2-15a* is very common in low-frequency amplifiers. There is no tuned circuit, so you are not dealing with an RF or IF amplifier. The disadvantage of this kind of coupling is that very low frequencies see a high reactance in the coupling capacitor. So the gain falls off rapidly at low frequencies.

Transformer coupling, shown in *Figure 2-15b*, is used throughout a wide range of amplifier frequencies. If the transformer is not tuned or if it has an iron core, as shown, you are definitely dealing with a low-frequency amplifier - usually in the audio range. On the other hand, if the transformer is tuned and has a ferrite or powdered iron core, you are dealing with a high-frequency IF or RF amplifier.

One version of impedance coupling is shown in *Figure 2-15c*. In modern circuits impedance coupling has become quite complex with computer-designed configurations.

The overall purpose of the circuit is to pass a band of frequencies. In that sense, it is similar to transformer coupling. In the circuit shown, the inductor is used to increase the high-frequency response of the coupling network. That, in turn, increases the range of low-frequency signals passed between the two amplifiers.

Impedance coupling is made with combinations of resistors, capacitors and inductors. If it is an RF circuit the capacitors and/or the inductors may be tunable in order to set the range of frequencies that can be passed.

Figure 2-15d shows direct coupling. The signal does not pass through any reactive components as it goes from the output of the first amplifier to the input of the second. That gives it a wide frequency range, but the system is not without its problems which will be discussed in the next section.

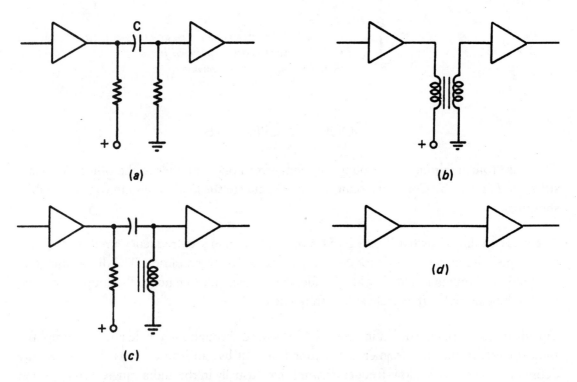

Figure 2-15. Four methods of coupling signals between circuits. a) R-C coupling. b) Transformer coupling. c) Impedance coupling (one example). d) Direct coupling.

LEVEL SHIFTING

Direct coupling requires a good power supply because the DC levels of the second amplifier must be higher than the DC levels of the first one. That causes a condition known as level shifting and it is illustrated in *Figure 2-16*. Note that the input DC voltage of the second amplifier has to be at the same DC level as the output level of the first amplifier. Therefore, the collector voltage of the second amplifier must be even higher.

It is not unusual to have four or five direct-coupled amplifiers in order to get a desired gain at a broad bandwidth. The problem is that the DC level of the signal is shifted in each amplifier and the power supply must be able to accommodate the increased DC levels in each stage.

The relationship between the signal and the DC level is shown in *Figure 2-16*. That signal offset is undesirable. It can be particularly compensated for by using the upside-down PNP transistor as the last stage. The PNP transistor returns the signal level to very-near common. At the same time, it produces additional gain in the chain of amplifiers.

Instead of the upside-down transistor, an NPN transistor emitter follower transistor circuit can be used. It will produce a broad bandwidth but has the disadvantage that its voltage gain is less than 1. Therefore, an amplifier is being used without any additional voltage gain. That may be undesirable since gain is an important factor in a direct-coupled circuit.

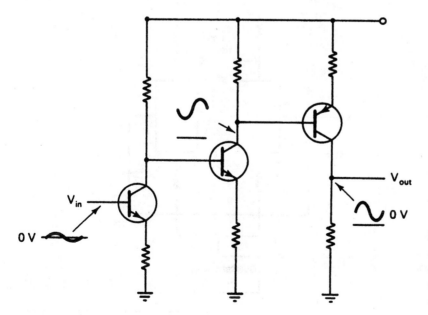

Figure 2-16. The problem of level shifting in direct-coupled amplifiers is partially reduced by using the PNP (upside-down) configuration.

DIFFERENTIAL AMPLIFIERS

A differential amplifier is shown in *Figure 2-17*. One way to use the circuit is to deliver two signals 180 degrees out of phase to inputs x and y. The output signal is proportional to the difference between the input signals. Since one signal is going positive while the other is going negative, their difference is greater than either signal taken by itself. That explains the high gain of the differential amplifier.

If inputs x and y are connected together it is called a common-mode connection. A signal delivered to the common-mode input should produce no output signal because the voltages at the output terminals rise and fall together. One way of testing differential amplifiers, then, is to look for a zero-volt output signal when the input is connected in a common mode. A signal is delivered to the common-mode connection in this test.

An important feature of a differential amplifier is the constant-current device that controls the amount of current into the two amplifiers. Because of that device, anything done to increase the current in one of the amplifiers will automatically decrease the current by the same amount in the other.

Differential amplifiers are often used for the input circuit of operational amplifiers. In *Figure 2-17*, long-tail bias is used, but some of the new operational amplifiers can be used with a single-ended power supply.

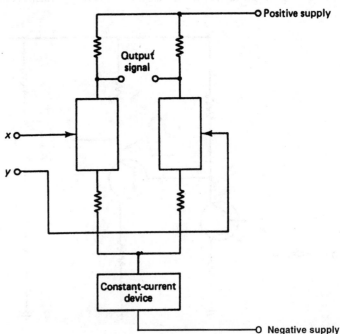

Figure 2-17. A differential amplifier has a very high gain. This type of amplifier is often used as the input circuit for integrated circuit amplifiers.

POWER AMPLIFIERS

There are some important characteristics of power amplifiers to watch for when troubleshooting those circuits. They must have a high input signal voltage, but the output signal voltage is usually not high. A power amplifier is used to convert signal voltage into signal current, and it is the current that produces the power in the output circuit.

You can expect a very low voltage gain in a power amplifier. Also, they are often operated with heat sinks that have cooling fins. The output of a power amplifier operates a device that requires high current, often some form of a transducer or motor. A loudspeaker is an example of such a transducer.

Figure 2-18 shows a single-ended power amplifier. The input signal is direct-coupled from the previous audio voltage amplifier stage. The emitter stage is unbypassed for a broader frequency response. The power transistor is normally mounted on a heat sink. The circuit is inefficient compared to other power amplifiers because the transistor is operated Class A. However, it is very popular in low-priced equipment.

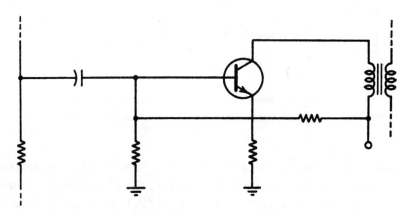

Figure 2-18. A single-ended power amplifier. This is not an efficient circuit.

Figure 2-19 shows a push-pull power amplifier. That design will produce more output power than a single-ended amplifier. It is designed in such a way that when the signal at x is going positive the signal at y is going negative. The transformer secondary splits the phase. The output is produced by first one amplifier and then the other to complete a cycle.

Because of the way they are designed, push-pull amplifiers must be operated Class AB, B, or C. In an audio amplifier, that is usually Class B or AB. Class AB amplifiers have a slight forward bias. It is not biased at cutoff like Class B amplifiers. The slight forward bias is used to overcome the 0.7V present between the emitter and base of bipolar transistors.

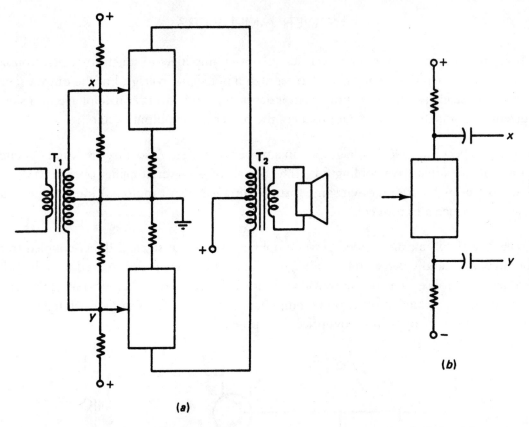

Figure 2-19. Power amplifiers in a push-pull configuration produce more output power than the single-ended type. a) A phase splitter is needed. In this case a transformer is used. b) Phase shifting can be accomplished by an amplifier, as shown here.

Without the slight forward bias, there is a period of time when neither amplifier is conducting. That produces an undesirable condition in the output called crossover distortion. It is illustrated in *Figure 2-20*. So, if the amplifier in *Figure 2-19a* is made with bipolar transistors, you can look for some kind of forward DC bias on the transistors.

If one of the two amplifiers in the push-pull circuit is defective, the other one may still produce a very distorted output.

The 180-degree out-of-phase signals for the push-pull amplifier of *Figure 2-19a* are produced by a transformer that is center tapped. The out-of-phase signals may also be produced by a phase splitter, as shown in *Figure 2-19b*. In that case, one output is taken from the DC output terminal and one is taken from the DC input terminal. That makes the signals at x and y 180 degrees out of phase, and those out-of-phase signals can be used to operate the two amplifiers in a push-pull circuit.

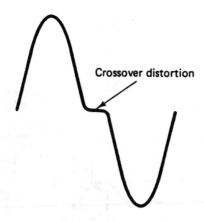

Figure 2-20. *Crossover distortion is primarily a problem with bipolar transistor power amplifiers.*

At one time, only vacuum tubes or bipolar transistors were used in the push-pull amplifier configuration. However, special MOSFETs, called VFETs, are capable of power amplification and they are now showing up as power amplifiers in various circuits. A popular power amplifier that does not require a phase splitter or output transformer is shown in *Figure 2-21*. It is called a complementary amplifier. That amplifier has sometimes been referred to as a *totem pole*. The circuit is presented in a basic form, but the essential parts are present.

A PNP and NPN transistor are used to alternately charge and discharge capacitor C. On the positive half of the cycle the NPN transistor is forward biased and the capacitor charges in the direction shown by the solid arrows. On the next half cycle the negative-going input signal cuts OFF the NPN transistor and forward biases the PNP type. That transistor conducts heavily and discharges the capacitor in a path shown by broken arrows.

Since the current through the speaker is in one direction during charging and the opposite direction during discharging, the speaker is actually carrying an amplified version of the AC signal delivered to the complementary amplifier.

The complementary power amplifier is very popular in operational amplifier output circuits and in many receiver systems. It does not need to be operated with bipolar transistors. Power-operating VFETs can also be used. However, a vacuum tube cannot be used in this configuration because there is no such thing as a complementary vacuum tube. They all operate with positive voltages on the plate with respect to the emitter.

Long-tail bias is shown in the circuit of Figure 2-21, but single-ended versions are available. The advantage of long-tail bias is that the output of the amplifier is at zero volts. You can actually short-circuit the amplifier output to ground without destroying this type of power amplifier circuit.

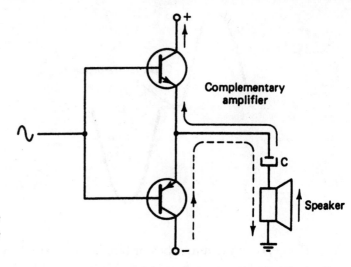

Figure 2-21. This is a simplified version of a complementary amplifier. Note that no phase splitter is required.

THE RECEIVER AS AN ELECTRONIC SYSTEM

As mentioned in the Preface, a simple AM receiver will be used to represent a system. That is not because this book is about repairing radios but, rather because that basic system is easy to understand and is familiar to most technicians.

The basic idea of the receiver is reviewed in block diagram form shown in *Figure 2-22*.

All receivers have four sections in common:

> There must be an antenna *reception* system to deliver the incoming signal to the *selection* part.

> There must be a tuner that permits the receiver to select one station and reject all others.

> There must be some kind of *detection* to separate the carrier wave and audio waveforms. The audio waveform is then delivered to the audio amplifier.

> There must be some method of *reproduction*. It is the speaker in the system illustrated.

Instead of using a converter - that is, a combination of an RF amplifier, a mixer, and an oscillator - it is possible to have them as separate stages. See *Figure 2-23*. In that case, the combination RF amplifier and oscillator results in a separate mixer stage.

In the circuits of *Figure 2-22* and *Figure 2-23* it is necessary to use a nonlinear converter stage to combine the signals.

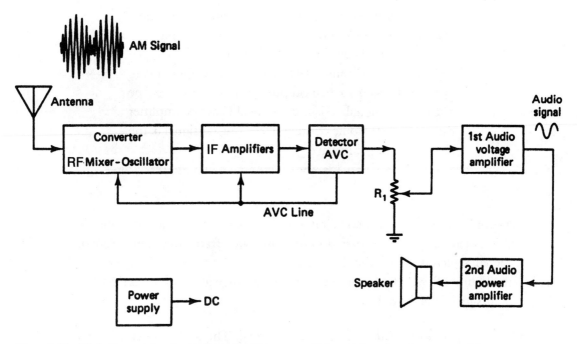

Figure 2-22. Block diagram of a simple AM radio. This system will be used throughout the text to explain system concepts.

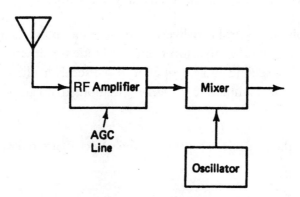

Figure 2-23. The front end of a receiver may consist of individual sections. When the sections are combined into one, the circuit is called a converter.

> **Note**
>
> You cannot heterodyne or modulate signals in a linear amplifier circuit. Nonlinear means that the amplifier is operated Class B or Class C or, in the case of the bipolar transistor versions, Class AB. Linear amplifiers are operated Class A so their output signal is the same shape as the input signal. Also, in an ideal Class A amplifier there is no harmonic content in the output signal if the input is a pure sine wave.

The oscillator frequency is tuned so that it is 455 kilohertz above the incoming signal. When the radio is tuned from station to station, the oscillator frequency is also tuned. The 455-kHz IF signal, which is always the difference between the RF and oscillator signal, is delivered to the IF amplifiers. The IF amplifiers provide a high amount of gain for the signal and their output signal is delivered to the detector.

The detector separates the audio from the IF signal. The audio signal is delivered to the audio amplifier stages through the volume control (marked R_1 in the block diagram). Note the AVC line that comes from the detector. The stronger the incoming signal, the higher the AVC voltage and the more negative feedback that occurs in the receiver. So, the AVC voltage controls the gain of the first stages in such a way that stronger signals result in less gain.

The output of the volume control is delivered first to an audio voltage amplifier. Its purpose is to increase the signal *voltage* to a sufficient amplitude for driving the power amplifier. That power amplifier is a second audio amplifier in the receiver. Its purpose is to convert the high-voltage input signal amplitude to *current* variations that can operate the transducer (speaker).

All of the stages (with the possible exception of the detector) are fed by a DC power supply voltage.

The block diagram of *Figure 2-22* represents the receiver shown in schematic form in *Figure 2-24*.

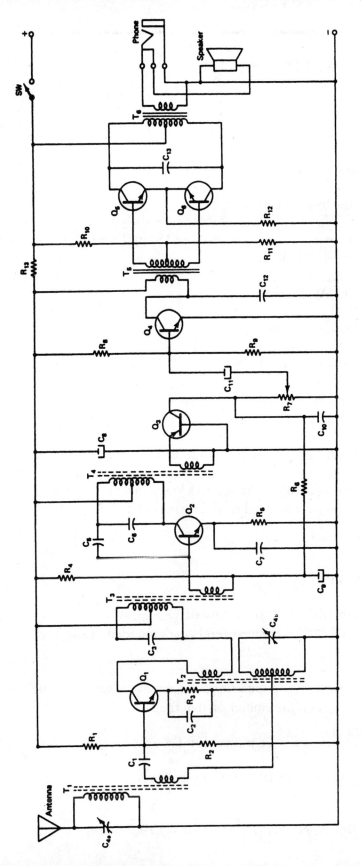

Figure 2-24. *Circuit for an AM radio. The purpose of every component is explained in the text.*

The following lists show the purpose of each component in the schematic of *Figure 2-24*.

RESISTORS

R_1 and R_2 - voltage divider bias for Q_1

R_3 - temperature stabilization resistor for Q_1

R_4 - part of the bias circuit for Q_2

R_5 - provides temperature stabilization for IF amplifier Q_2. It also provides negative feedback for Q_2.

R_6 - part of the AVC filter

R_7 - receiver volume control

R_8 and R_9 - voltage divider for biasing Q_4

R_{10} and R_{11} - provide voltage divider bias for the push-pull amplifier. This circuit provides Class AB operation.

R_{12} - emitter temperature stabilization resistor for the push-pull amplifier

R_{13} - acts to drop the supply voltage to a lower value for the voltage amplifiers; it is also part of a decoupling filter comprised of R_{13} and C_8

CAPACITORS

C_1 - couples the RF signal to the base of Q_1 and, at the same time prevents the base bias of that transistor from being grounded through windings on T_1 and T_2

C_2 - emitter-resistor bypass capacitor; it increases the gain of Q_1 by eliminating signal voltages on the emitter of that transistor

C_3 - tuning capacitor for the primary of T_3

C_{4a} and C_{4b} - RF and oscillator tuning capacitors

C_5 - feedback capacitor to increase the high-frequency gain of Q_2. Note that it is connected to the primary of T_4 in such a way that it provides a regenerative feedback.

C_6 - tuning capacitor for the primary of T_4

C_7 - emitter-resistor bypass capacitor used to prevent degeneration and thereby increase the gain of Q_2

C_8 - part of the voltage decoupling filter (see also R_{13})

C_9 and C_{10} - along with R_6. This is the AGC (AVC) filter.

C_{11} - coupling capacitor between the volume control and the audio voltage amplifier

C_{12} - this low-value capacitor removes any residual IF signal that gets through the system

C_{13} - false bass tone control. High audio frequencies are eliminated so as to give the output a more bass sound.

TRANSFORMERS

T_1 - RF transformer that is part of the circuit that selects one of many stations

T_2 - oscillator transformer - it provides tuning for the oscillator signal and regenerative feedback for the oscillator signal.

T_3 - first IF transformer

T_4 - second IF transformer

T_5 - phase splitter for the audio push-pull amplifier

T_6 - output transformer for the push-pull audio amplifier

TRANSISTORS

Q_1 - converter

Q_2 - IF amplifier

Q_3 - detector - this transistor is used as a diode detector.

Q_4 - audio voltage amplifier

Q_5 and Q_6 - push-pull audio power amplifiers

FINDING TROUBLE IN A DEAD RECEIVER

A good way to start troubleshooting any system is to mentally divide the system in half. Using the example of a radio again, the volume control is an easy place to start. If you inject an audio-frequency signal at the top of the volume control you should be able to hear the audio at the speaker. You may have to adjust the volume control for sound.

If you hear the sound, you can assume the circuits between the volume control and output are OK. If not, move the test signal forward (toward the volume control) point by point, until you hear the sound. When you do, you have just passed the point where the trouble is located. If you do hear the sound when you inject the signal at the volume control, then it follows that the trouble is somewhere between the volume control and the antenna.

You have to change to a modulated RF signal and inject that signal into the first IF amplifier. In an AM radio you would use 455 kHz for the IF frequencies and any frequency from 550 to 1600 kHz at the antenna. Trace through the radio from the volume control to the antenna using the proper signal.

The preceding method works very well. You can add the use of an oscilloscope to observe the waveforms as they pass through various sections. You can use a sweep generator to look at the bandpass of the IF transformers and amplifiers. This is good experience, but it is also a good way to go broke. This is *not* saying that test equipment is not important and useful.

Tapping the center tap of the volume control with a screwdriver is a quick way to inject a signal. The tapping should produce a clicking noise in the speaker. If it does, then the audio system is probably OK and your time would be better spent looking at the sections in front of the volume control. After looking for obvious faults and measuring the power supply voltage, a few quick checks with a signal generator should be used to determine where the fault lies.

OPERATIONAL AMPLIFIERS

In mathematics an operator is a symbol that tells you what to do. For example, a plus sign indicates that you are to add the second number to the first. Likewise, a division sign means that you are to divide the first number by the second.

Operational amplifiers (Op Amps) were used in early analog computers because they were used to perform arithmetic operations. A long time ago it was learned that it was unnecessary to design a new amplifier for every new arithmetic operation. If the gain of an amplifier is sufficiently high, its characteristics are almost completely controlled by the type of feedback that is used with that amplifier.

Figure 2-25 shows two typical operational amplifier circuits. Power supply connections are not usually shown with op amp symbols. The + and - signs refer to the signal noninverting and inverting input terminals. Note that a signal delivered to the inverting input results in an output signal that is 180 degrees out of phase with the input.

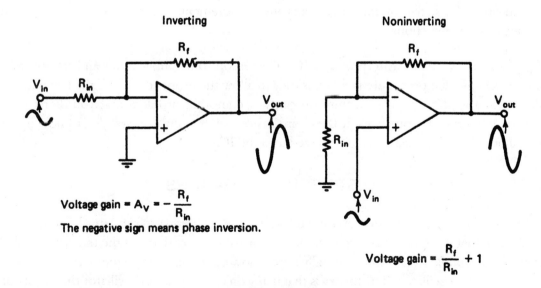

Figure 2-25. Two examples of operational amplifiers: inverting and noninverting.

There are some important characteristics that you should know about an operational amplifier in terms of troubleshooting. The voltage at the inverting input terminal is maintained at zero volts. The amplifier will do anything that it has to in order to keep that DC voltage at zero volts.

If the feedback resistance (R_f) is high, then the amplifier must have a high gain in order to overcome the feedback signal opposition. Remember that the feedback signal is used to maintain the input DC voltage at zero volts. If the feedback resistance is low, then the gain is low because not much output signal is needed to feed back to the common mode input in order to maintain the voltage at zero volts.

Operational amplifiers found a new lease on life when they became integrated circuits. That allowed them to be used for a very wide variety of applications. You will see operational amplifiers in every type of electronic system today.

Op amps are highly reliable and require little troubleshooting. They are typically low-frequency amplifiers, so an oscilloscope check of the input and output signals is the usual procedure. Look for distortion, especially clipping, in the output signals. For many years the op amps used only long-tail (positive/negative) bias. Now there are single-ended op amps. As with any integrated circuit (IC) device, check the power supply inputs.

The next statement is controversial so use your own judgment: When you check any integrated circuit for signals or voltages, make your measurements at the pins, not at the printed circuit board connections.

Figure 2-26 shows that way to probe an IC. The reason some publications and schools take a stand against that procedure is that it is easy to allow the probe to slip off the terminal. If the probe shorts between two pins you can cause another problem that you didn't have before. However, if you check at the board instead of the pin, you still don't know if the signal or voltage is actually being delivered into the IC.

TESTS FOR TRANSISTORS

When a transistor is defective in a circuit voltage, measurements can usually be used to determine that it is bad. Furthermore, there are in-situ testers that test the transistor *while it is wired into a circuit.* In-situ testers usually apply a low operating voltage to the transistor and operate it as an oscillator. The theory is that if it won't operate as an oscillator the transistor must be defective.

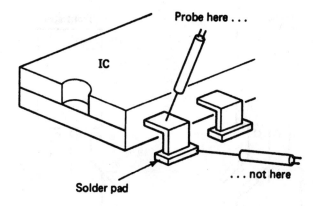

Figure 2-26. In the opinion of the authors, this is the correct way to probe signals and voltages on an integrated circuit.

Remember that an oscillator is nothing more than a regenerative feedback amplifier circuit that has a very high gain and a very narrow bandwidth.

When you are replacing a transistor, it is desirable to test the new one before it is soldered into place. The very best testers are those that check the beta of the transistor. Remember that the beta of a transistor is determined as follows:

$$\beta_{DC} = \frac{I_C}{I_B} (I_E \text{ held constant})$$

$$\beta_{AC} = \frac{\Delta i_c}{\Delta i_B} (i_E \text{ held constant})$$

Ohmmeters can be used to check bipolar transistors and other components. However, as a general rule, the ohmmeter tests are not as reliable as beta tests or dynamic operating tests. Nevertheless, you may be in a situation where you do not have a beta checker available. Certainly the ohmmeter test is better than nothing.

The ohmmeter test is illustrated in *Figure 2-27*. In order to perform this test, it would be best if you know which of your ohmmeter leads is positive and which is negative. In the illustration, it is assumed that the lead with the alligator clip is the negative side of the ohmmeter. Remember that an ohmmeter supplies a voltage to the circuit being tested.

You should never test a transistor, or any other semiconductor device, when the ohmmeter is on the X1 (times one) position. In that position many ohmmeters are capable of supplying high current that can destroy a semiconductor device. So, for this test, it is assumed that the ohmmeter is not on the X1 position. The X100 scale is recommended.

Troubleshooting Basic Analog Circuits 73

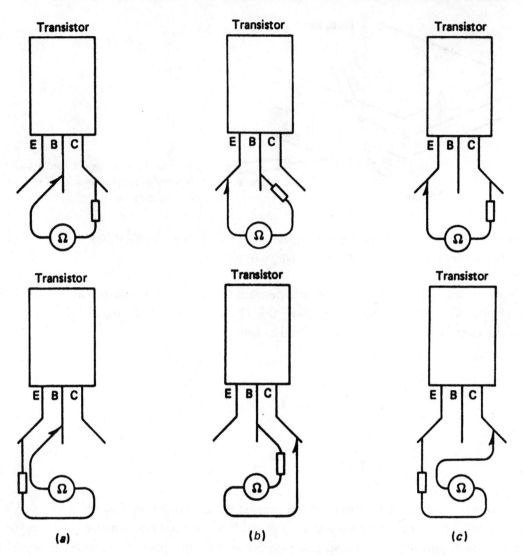

Figure 2-27. Use of an ohmmeter to determine if a bipolar transistor is capable of performing its function.

In *Figure 2-27a*, the negative lead of the transistor is connected to the base. The positive lead is then connected first to the collector and then to the emitter. If you are testing a PNP transistor, the connection in will forward bias both of the junctions. If it is an NPN transistor, it will cause a reverse bias. So, the transistor can be identified as being either an NPN or PNP type.

Suppose it is the PNP type. Then both of the connections in *Figure 2-27a* should show a relatively low resistance. On the other hand, if it is an NPN type, both of the connections should show a high resistance.

Changing to the positive lead on the base, as shown in *Figure 2-27b*, should reverse the conditions that you got in *Figure 2-27a*. So, if you got low resistance in both directions in *Figure 2-27a*, you should get a high resistance in both directions in *Figure 2-27b*. On the other hand, if you got high resistances in *Figure 2-27a*, you should get low resistances in *Figure 2-27b*.

You have checked the emitter-base and collector-base junctions to see if they are capable of conducting or rectifying.

A third test should also be used. It is illustrated in *Figure 2-27c*. Now the ohmmeter is connected from emitter to collector. You should measure a high resistance in both connections shown. The base lead is open, so the transistor is not forward biased. That, in turn, means a low emitter-collector current. This particular ohmmeter test is essential for power transistors that sometimes have a shorted emitter-to-collector connection.

Some technicians prefer the test illustrated in *Figure 2-28*. It is easy to make a simplified test jig where the transistor can be plugged in and the ohmmeter used for the power supply.

With the switch open, the ohmmeter should show a high resistance. When the switch is closed, the transistor is forward biased and the transistor should have a low resistance. This test is better than the ohmmeter test because it is dynamic - that is, it tests the transistor's ability to a control current flow by changing the amount of base current. However, it is still not as good as the beta test.

In any case, the judgment about which test is better depends on the preference of the technician and the types of troubleshooting the technician is doing.

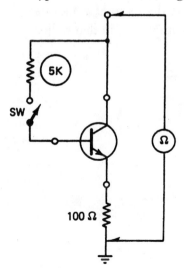

Figure 2-28. A better way to check a bipolar transistor with an ohmmeter.

Circuits That Could Be Represented by Logic Gates

Remember that there are only two possible levels for any lead of a logic gate: 1 or 0. You can allow one set of conditions such as the presence of a signal, to be represented by a 1, and another set of conditions, such as a no signal, to be represented by 0. In each of the test questions you should use your knowledge of basic gates to determine how the circuit should be represented by logic gates.

Summary of Linear Circuits

There are some basic voltages and polarities that a technician *must* know when troubleshooting. Technicians do not have time to search for the voltages on schematic diagrams or in books when they are looking for the cause of a problem. It is something that they must know in order to troubleshoot efficiently. The voltages and polarities have been summarized in this chapter. There has also been some discussion on basic circuit configurations that often require troubleshooting.

A logical question at this time is: How is this troubleshooting information affected by the use of integrated circuits in some modern systems? The answer is that with integrated circuits, troubleshooting is almost entirely on a system basis. So the techniques of signal tracing and signal injection are most important for those systems. Always keep in mind, however, that even integrated circuit systems have outboard transistor amplifiers and transistors used for specialized purposes. You will need the information in this chapter for troubleshooting those types of circuits.

Because it provides a basis for troubleshooting discussions, a radio system has been reviewed briefly, as well as the purpose of each component in that radio.

Chapter 2 Quiz

1. The amplifier shown in *Figure 2-29*

    A. has a gain less than 1.

    B. can be used for level shifting.

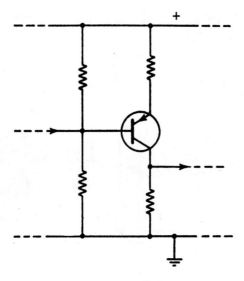

Figure 2-29.

2. Which of the following types of solid state amplifying devices might have +450V on its DC output electrode?

    A. Depletion MOSFET

    B. Enhancement MOSFET

3. Can an enhancement MOSFET amplifier be self-biased with a source resistor?

4. Refer again to *Figure 2-24*. If R_9 is open, then transistor Q_4 will be

    A. saturated.

    B. cut off.

5. Refer again to the system in *Figure 2-24*. What is the purpose of C_9, R_6 and C_{10}?

    A. It is a filter.

    B. It is an RC coupling circuit.

6. Refer to *Figure 2-30*. The polarity of the power supply voltage at 'Y' should be

    A. positive.

    B. negative.

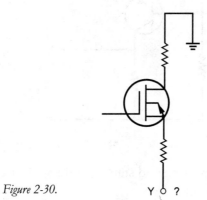

Figure 2-30.

7. Refer to *Figure 2-31*. The polarity of the gate voltage at 'X' should be

    A. positive with respect to common.

    B. negative with respect to common.

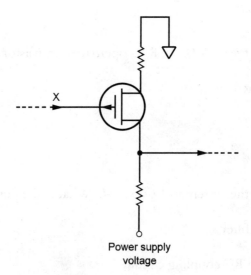

Figure 2-31.

8. In the complementary connection of *Figure 2-32*, if point 'X' is accidentally shorted to common it will

    A. destroy Q_1.

    B. destroy Q_2.

    C. not destroy either transistor.

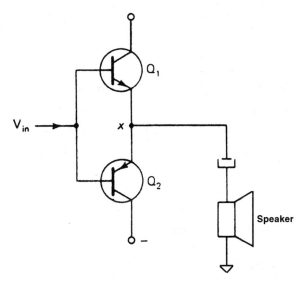

Figure 2-32. Note the slight forward bias on the bases is not shown.

9. In a complementary power amplifier, like the audio output in *Figure 2-32*, the transistors are slightly forward biased. In other words, they are operated Class AB. For what is that forward bias used?

10. Which type of field effect transistor is biased like a PNP transistor?

11. In a Class A amplifier made with an NPN transistor you would expect the DC base voltage to be

    A. positive with respect to the collector.

    B. negative with respect to the collector.

12. Which types of field effect transistors have the same polarity of voltage on their electrodes as vacuum tubes?

13. Increasing the gain of an amplifier will automatically

    A. increase its bandwidth.

    B. decrease its bandwidth.

14. What is the first measurement made with test equipment when troubleshooting?

15. What type of diode is used as a capacitor?

16. To reduce base-collector capacity, the collector voltage is

    A. increased with respect to the base voltage.

    B. decreased with respect to the base voltage.

17. In a tetrode vacuum tube the screen grid voltage should be

    A. positive with respect to the cathode.

    B. negative with respect to the cathode.

18. Which amplifier configuration has a low input impedance and a high output impedance?

    A. Common gate

    B. Common drain

19. Which of the amplifying devices discussed in this chapter can be used with either AGC or power supply bias?

20. Which of the following is used to increase the bandwidth of an amplifier?

    A. Regenerative feedback

    B. Degenerative feedback

21. Which logic gate has the truth table in *Figure 2-33*?

A	B	L
0	0	0
0	1	1
1	0	1
1	1	0

Figure 2-33.

22. Which type of gate is represented by the symbol in *Figure 2-34*?

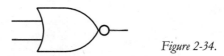

Figure 2-34.

CHAPTER THREE

Troubleshooting with Meters and Logic Probes

Overview

A certain amount of personal opinion is involved in troubleshooting methods. One technician may prefer to use a voltmeter for most troubleshooting problems. Another technician will always reach for the oscilloscope leads.

Even though you may have your own preference, you should be familiar with all of the methods of troubleshooting. This chapter deals with the voltmeter as a troubleshooting instrument. The digital probe is also discussed for troubleshooting digital circuits.

Both analog and digital multimeters [volt-ohm-milliameters (VOMs)] are available for troubleshooting analog circuits. Although you may use only one type, you should know the advantage and disadvantages of each. Also, you should know their limitations.

Objectives

How is the input resistance of an analog volt-ohm-milliammeter estimated?

How do you make use of a meter turnover?

How do you get maximum accuracy with analog VOM measurements?

What are some methods of testing transistors and thyristors with a VOM?

How do you avoid misuse of the dB scale on a VOM?

What is a quick method of estimating voltages?

How do you know when a voltmeter test won't work?

How are logic gates combined to make digital circuits?

How do you determine all of the possible inputs of a combinational logic gate circuit?

How are equations for combinational logic circuits determined?

What does a truth table tell you?

ANALOG

Here are a few precautions regarding the use of a VOM:

Don't measure voltage across a high-resistance or high-impedance circuit with a meter that has a relatively low input resistance.

Don't try to measure nonsinusoidal voltages with a VOM. Most VOMs are calibrated to display only sine wave voltages.

Don't use a meter to measure in MOSFET circuits and integrated circuits with MOSFETs unless you know the probes are static-free. Newer MOSFET circuits are buffered internally with zener diodes to prevent destruction by static charges. However, there are still some of the older types around. Just to be sure you are on the safe side, make sure your probes don't carry a static charge.

Don't use a VOM to measure the open-circuit voltage across a cell or battery because it won't tell you anything useful about its condition. This problem is discussed in detail in the chapter.

Don't use a lot of test equipment for measurements when a basic VOM or scope can do the job.

Don't use a voltmeter (instead of a logic probe) to measure logic 1 and logic 0 in a digital circuit.

ABOUT THE METER

In addition to their ability to measure voltage, current and resistance, there are other important features of meters used in modern servicing.

In analog meters there is a choice between jeweled and taut-band meter movements. One important difference is the way the return force is applied to the pointer. The return force is needed to return the pointer to zero and to provide the necessary opposition as the pointer moves upscale. *Figure 3-1* shows the concepts. The taut-band movement in analog meters is the more rugged of the two and will hold its accuracy over a longer period of time.

Some analog meters have a mirror on the scale to prevent parallax. Parallax error is a

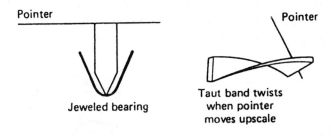

Figure 3-1. The basic idea behind jeweled and taut-band meter movements.

problem in measurement accuracy. It occurs when you view the pointer at an angle. An example of parallax is when a car passenger looks at the speedometer pointer from an angle. Since the passenger is sighting along the pointer and speed scale at an angle, it appears that the speed of the car is lower than it really is.

A mirror on the scale can be used to eliminate parallax error. The procedure is to view the meter with one eye when the pointer is up-scale. When you are looking at it in a position where you cannot see the reflection of the pointer in the mirror, then you are looking at the meter movement from a vertical (most accurate) position.

A low-power ohms scale (on some meters) is needed to make accurate resistance measurements in semiconductor circuits. As shown in *Figure 3-2a*, the ohmmeter leads have a voltage difference that can forward bias a semiconductor junction. In the case shown, that would put R_1 in parallel with R_2 and the junction resistance.

If you don't have a low-power ohms scale, you can estimate the parallel resistance of R_1 and R_2 and disregard the resistance of the forward-biased transistor. Using jumper leads, place an emitter-to-base short circuit on the transistor as shown in *Figure 3-2b*. Then, measure the parallel resistance of R_1 and R_2.

The measurement in *Figure 3-2b* will accurately tell you if either resistor is open or greatly changed in value.

Another way is to reverse the leads of the ohmmeter to prevent the junction from being forward biased. You have to be careful not to reverse bias a semiconductor junction by a great amount. It is not necessarily the best way to make the measurement.

You have to consider *working quickly* to be a very important factor when measuring with a meter, scope or any other measuring device. Technicians who work fastest clamp the common lead of the meter onto a circuit common point with an alligator clip and make all measurements with a single probe. That eliminates a lot of wasted time that would occur when you move both probes around the circuit.

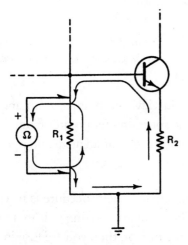

Figure 3-2a. In this measurement the ohmmeter is forward biasing the emitter-base junction of the transistor. That produces a parallel path for ohmmeter current through resistors R_1 and R_2. Obviously, the ohmmeter will not correctly display the resistance of R_1.

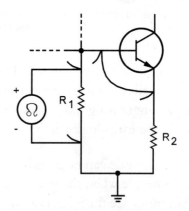

Figure 3-2b. With the jumper connector between the emitter and base the parallel resistance of R1 and R2 is accurately known.

With a jumper across the emitter-base junction, measure across R_1 (or R_2). The two resistors are in parallel. You can estimate the parallel resistance by using the following equation:

$$\text{Resistance}(R_{eq}) = \frac{R_1 R_2}{R_1 + R_2}$$

Use of the low-power ohms scale is preferable if you have that provision on your VOM or DVM (Digital VoltMeter). It does not supply enough voltage to turn the junction ON. So R_1 in *Figure 3-2b* is no longer in parallel with R_2, and the resistance of each resistor can be measured directly.

Analog meters are still very useful whenever a high input meter impedance is not absolutely necessary. One advantage of the analog meter scale is that it is easy to use in applications where adjustments require a null or a peak.

If you are using an older VOM, make sure that there are no static charges on the probes because those charges can destroy some types of integrated circuits (especially early CMOS types).

Vacuum tube voltmeters are still being used. They are analog meters with a very high input resistance (typically 20 megohms). There is nothing wrong with using a vacuum tube voltmeter, assuming there are no static charges on the probes; and, it still provides accurate measurements.

MEASURING AC CURRENT

Traditionally, analog meters and oscilloscopes do not have direct AC current-measuring capability. That does not mean you can't measure AC current with them, but it requires a special test setup.

Figure 3-3 shows how it is done. A 1-ohm resistor is connected in series with the circuit current to be measured. The AC voltmeter or oscilloscope is used to measure the voltage across that resistor. The current is numerically equal to the voltage when the inserted resistance equals one ohm:

$$I = \frac{V}{R} = \frac{V}{1}$$

so: $I = V$

Of course, the 1-ohm resistor must have a power rating high enough so it is not destroyed by the measurement.

$$\text{Power (P)} = I^2R = 1^2 \times R = R$$

A good way to measure AC current is to remove the fuse in the AC line and insert the 1-ohm resistor across the fuse holder connections using alligator clips.

If you insert a 10-ohm resistor, the voltage measurement must be divided by 10. As long as the resistance of the inserted resistor is less than 10% of the total resistance of the circuit, the method just described will give a very good estimate of the AC current.

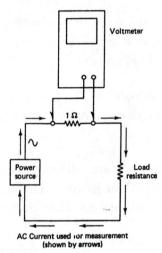

Figure 3-3. A 1-ohm resistor can be used in the measurement of AC current. This illustration shows the procedure.

OHMS PER VOLT

Analog meter movements are often rated according to their sensitivity, which is stated as the number of ohms per volt. A frequently-missed question in tests is given here:

Question: A certain meter is rated at 10,000 ohms per volt. When it is being used to measure 5V, on the 5V scale, its resistance is:

$$10,000 \frac{\text{Ohms}}{\text{volt}} - 5 \text{ volts} = 50,000 \text{ ohms}$$

This is incorrect. You can determine the *approximate* resistance of the meter by using the ohms-per-volt rating and the maximum (full-scale) voltage on a specific voltage range. *Figure 3-4* shows the calculation.

So, you can tell how much resistance the meter is offering when making a voltage measurement. For all practical purposes, that resistance is the same as the resistance of the multiplier, but that is not always known. The meter movement resistance is often disregarded in the calculation.

Example: A certain voltmeter with a 0-10V scale and a meter sensitivity of 20,000 ohms per volt is being used to measure a circuit voltage of 3V. The meter movement resistance is 60 ohms. How much resistance does the meter offer? In other words, what is the input resistance to the voltmeter?

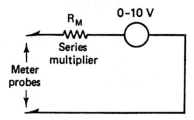

Approximate meter resistance = ohms per volt X full-scale voltage.

Figure 3-4. The approximate meter resistance can be obtained by this simple method.

Solution: Use the approximate method shown in *Figure 3-4*.

Meter resistance » full scale voltage x ohms per volt » 10 x 20,000

Meter resistance » 200,000 ohms (approximately)

The meter resistance for a given scale is the same regardless of the particular voltage being measured on that scale. So whether you are measuring 3V or 8V on the 10V scale, the measurement has nothing to do with the meter resistance *on that scale*.

METER LOADING

When a voltmeter is connected across a resistor for making a voltage reading there is always a certain amount of meter loading. That means the meter reduces the circuit resistance and the measured voltage will be less than the actual voltage without the meter.

If the meter resistance is at least ten times the resistance of the resistor that it is connected across you can assume that the measurement is a good approximation.

CALIBRATION

It is useless to troubleshoot with a VOM if that meter is not accurate. It is important that you periodically check the accuracy of your meters. That can be done in a number of different ways.

To check the accuracy of your voltmeter, connect it in parallel with one that is known to be good. The test is shown in *Figure 3-5*. If you have a regulated adjustable power supply, it is useful to check the meter scale at a number of different voltage values. For a quick check, connect the meter across a new 1-1/2V carbon-zinc dry cell (LaClanche type) that is known to be good. That is a reasonably accurate 1.5V source.

Another place to get an accurate voltage is the +3.6V or +5V supply for integrated circuits. In most cases that voltage must be maintained accurately. For a quick check of AC voltage measurements you can use a 6.3V filament transformer (or transformer with a higher secondary voltage). A variac is also very useful. Use the test shown in *Figure 3-5* with a Variac replacing the battery.

Current-reading meters are checked by connecting them in series, as shown in *Figure 3-6*. Since the same current flows through both meters, they should indicate the same current value.

Precision resistors are a good way to check the resistance ranges of an ohmmeter. Use one percent resistors for the check.

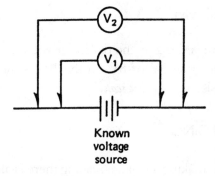

Figure 3-5. *This is the method used for checking the accuracy of a voltmeter against one known to be accurate.*

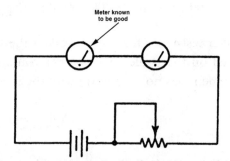

Figure 3-6. *This is the method of checking the accuracy of a current meter against one that is known to be correct.*

Most voltmeters have scales calibrated to display the RMS value of a sine wave voltage. In the case of an analog VOM with a Permanent Magnet, Moving Coil movement, the meter deflection is proportional to the average value of the voltage being measured. However, the scale is calibrated in RMS volts. That is based on the basic relationship:

$$V = V(avg) \times 1.11$$

For a more accurate value use:

$$V(RMS) = V(AVE) \times \frac{[1 \div \sqrt{2}]}{[1 \div \pi]} = 1.11072073$$

Where 1.11 is the form factor of a sine wave.

The equations given are only used for calibrating the AC voltage for the voltage scale! Remember, that relationship is only true for pure sine wave voltages. So the scale reading is incorrect if you are measuring a nonsinusoidal voltage.

You can use an AC meter calibrated for RMS value to measure the average value of a nonsinusoidal waveform if you know the form factor of the waveform. By definition:

$$\text{FORM FACTOR} = \frac{\text{RMS VALUE}}{\text{AVERAGE VALUE}}$$

so

$$\text{AVERAGE VALUE} = \frac{\text{RMS VALUE (SHOWN ON METER)}}{\text{FORM FACTOR}}$$

Question: A voltmeter shows a measured value of a sine wave voltage to be 10V. What is the average value being measured?

Answer: Average Value = $\dfrac{10}{1.110720737}$ = 9V

The form factor used here is well beyond a practical value. Normally, you would use 1.11. The meter may have an input circuit that changes the waveform slightly. So when the waveform is nonsinusoidal, the traditional way of making a measurement is to use an oscilloscope to measure the peak-to-peak value. That is the value you will see most often on schematics.

There is a trick used with VOMs to determine the nature of the nonsinusoidal waveform being measured. The trick employs the use of meter turnover. The procedure is to measure a sine wave AC voltage with the probes in one direction, then reverse the probes (see *Figure 3-7*). If you get the same reading both ways, you know that the full-cycle average value of the waveform is zero.

The example of turnover measurement just made can be useful in setting zero offset on a function generator. *Figure 3-8* shows how the offset adjustment affects the signal.

If you apply a signal that doesn't have zero offset to a differential amplifier, like the amplifiers you find at the input of an operational amplifier, the DC level will cause an unbalance. That, in turn, will usually result in distortion of the output signal.

Use the turnover measurement just described and adjust the signal generator offset until there is no turnover. Those quick tests will also give you an idea if the meter is working properly, but they are not sufficiently accurate for calibrating a meter.

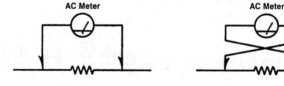

Figure 3-7. This procedure is used to ensure that the full-cycle average of an AC waveform is zero.

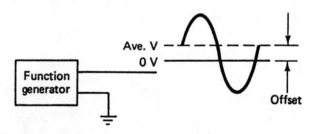

Figure 3-8. Function generators often have a DC offset adjustment. The waveform shown here indicates that DC offset is present. The AC meter measurement in Figure 3-7 demonstrates the DC offset.

Taut-Band and Jeweled Analog Meters

Taut-band instruments have two very important advantages. One is that they are more rugged than the jeweled type. The other is that there is no problem with stiction.

In the jeweled type, a jewel bearing is used to hold the pointer and a spiral spring is used to provide the restoring force. The problem is that there is some amount of starting friction, sometimes called stiction, between the pointer and the cup that it rides in.

Think of stiction as being a sticking friction. When a very slight change is required, as in measuring a low voltage, the jewel, especially in older instruments, tends to stick until a sufficient force is applied to start it. That is why you are advised to tap the meter gently when using it to make very small changes and adjustments.

To reduce the problem of stiction in a meter with a jeweled movement, the meter should always be put on its back so that the pointer is centered in the cup. Note in *Figure 3-9* that when the meter is standing the pointer rides down on the cup. That is especially true for older meters. It causes wear on the jeweled bearing, and it will cause inaccuracies in measurements.

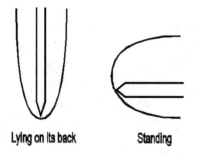

Lying on its back   **Standing**

Figure 3-9. The accuracy of measurement for jeweled meter movement is affected by the position of the meter during measurement. Note that when the meter is standing the pivot point is moved down to the side of the jeweled bearing.

Accuracy of Voltmeter Measurement

Meter manufacturers state that maximum accuracy occurs when the meter is reading full scale. Suppose, for example, the meter is being used to measure on the 10V scale and the manufacturer says that it will give a 2% accurate reading on that scale. That means it will be 2% accurate *only if you are reading 10 volts*. Usually, when the measurement is less than full scale the accuracy decreases.

That is why you are told to read analog meters when they are deflecting to some value in the upper one-third of the range for best accuracy.

METER PROBES

In addition to direct probes, there are two important types of probes used with voltmeters: high-voltage probes and demodulator probes. They are also used with oscilloscopes.

High-voltage probes act as a voltage divider to drop most of the high voltage being measured. That type of probe circuit is shown in *Figure 3-10*. If the probe is designed specifically for the meter, you simply connect it into the voltage terminal of the meter and read the voltage by using the indicated multiplying factor on the probe.

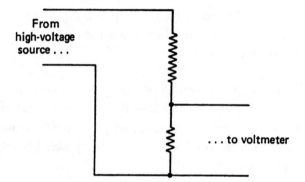

Figure 3-10. There is nothing more complicated in a high-voltage probe than a basic resistor voltage divider.

If you are designing your own series multiplier, be sure to use a full-scale voltage value for the highest voltage range. Determine the current that flows through the meter using the reciprocal of the ohms-per-volt scale. Then calculate your series resistor so that it can be used for a voltage at least 1-1/3 times the value of the highest voltage that you expect to measure accurately with that meter when using the multiplier. *Figure 3-11* shows a typical high-voltage probe.

A demodulator probe circuit is shown in *Figure 3-12*. That probe is useful for making voltage measurements in very high-frequency circuits. For example, if you want to measure the voltage in a television IF stage your meter may not be able to handle the 4-mHz measurement. However, with a demodulator probe you are removing the RF and looking only at the waveform envelope. Your meter will surely be able to give you some kind of indication.

Figure 3-11. High-voltage probes are made many different ways. This is an example of a self-contained high-voltage probe.

Note

Unless the modulation is a pure sine wave you will get an indication, but it will not be an accurate measurement of the modulated waveform RMS value.

Sometimes you are simply using the meter to see if there is an AC signal voltage present. Make sure there is a series capacitor so that DC in the circuit will not produce an upscale reading when you are checking for AC only. Some VOMs have a built-in isolation capacitor.

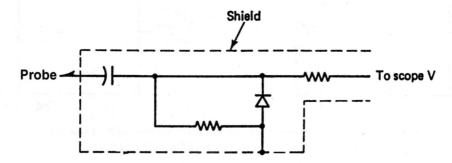

Figure 3-12. The demodulator probe shown here is easily constructed. It is very useful for measurements in RF circuits.

BATTERY TERMINAL VOLTAGE

This is a good place to repeat the first steps in troubleshooting:

> After preliminary visual tests and symptom analysis, measure the power supply voltage.

Almost any voltmeter will present a sufficiently-high resistance so that it does not draw enough current to properly load a battery for testing.

Figure 3-13 shows a voltmeter being used to measure the terminal voltage of a dry cell. Assume the dry cell is very old and has a high internal resistance. Under normal load the terminal voltage of the battery drops to a value that is too low to make the battery useful. However, measuring the terminal voltage with a voltmeter produces such a low value of voltage drop across the internal resistance that the terminal voltage seems to indicate that the battery is good.

So, a battery should always be tested under load. If you are troubleshooting an electronic system that uses battery power, the best load is to use the circuit the battery is supposed to supply power for. In other words, measure the battery voltage with the system ON. Reject a battery that has an under-load terminal voltage less than 80% of the full rated voltage. (The 80% value is an average. It may differ from some specific applications.)

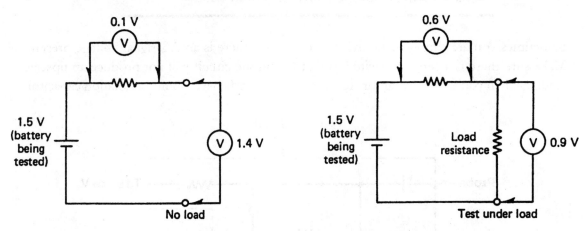

Figure 3-13. This illustration shows why the condition of a dry cell cannot be determined by an open terminal measurement. The measurement must be made with the call under load, as shown in the right-hand drawing.

Additional Information on Ohmmeter Testing

An obvious use of the ohmmeter scale on a digital multimeter (DMM) is for measuring the resistance of resistors. Some early digital multimeters had poor accuracy on the ohms scale. Newer designs, however, can now be used for very accurate measurements.

Caution

Never use the X1 scale on an ohmmeter when you are checking a pn junction! Some meters will supply junction-destructive currents on the X1 scale.

Figure 3-14 shows how an ohmmeter can be used to evaluate the condition and type of bipolar transistor. This isn't a test that you would run very many times before you get tired of it. If you do a lot of transistor testing, it is better to buy or build a beta checker, or use the beta checker that is available on many of the new digital multimeters. If a beta tester is not available a test setup can be hard-wired. However, remember that it checks the transistor for its ability to control current. It does not check for gain.

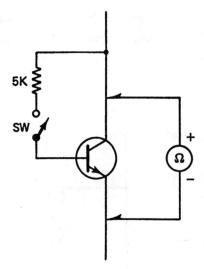

Figure 3-14. The ohmmeter resistance should be high with SW open and lower with SW closed. Reverse ohmmeter leads for PNP test.

Two ohmmeter tests for vacuum tubes are shown in *Figure 3-15*. The tests shown are for open- or closed-filament and low-resistance filament-to-cathode shorts.

Other tests show whether electrodes are directly shorted. For example, in a high amplification factor tube the grid might become shorted to the cathode. As with the transistor test, that is a once-in-a-while quick test. Tube testers do a much better job.

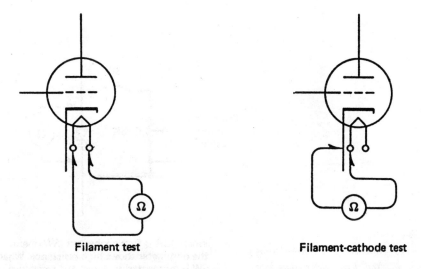

Figure 3-15. Here are two important ohmmeter checks for vacuum tubes. A third possible check is to measure the resistance between the grid and cathode.

Field effect transistors can be tested with an ohmmeter as shown in *Figure 3-16*. Commercial testers do a much better job of testing FETs. The ohmmeter quick test for thyristors is shown in *Figure 3-17*.

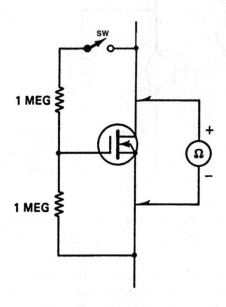

Figure 3-16. The MOSFET check shown can be used for other types of field effect transistors.

Ohmmeter should show decrease in resistance when switch is closed.

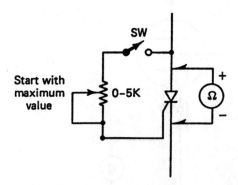

Figure 3-17. Thyristors can be checked using this technique. You should always start with the resistor adjusted to its highest value.

When SCR is first connected (SW open), the ohmmeter shows high resistance. When SW is momentarily closed, the resistance drops and remains at the lower value.

Testing Capacitors with a VOM

The condition of capacitors is difficult to determine with an ohmmeter. Usually, an ohmmeter does not deliver more than a 1-1/2 volt output. So, when you connect the leads across a capacitor it may not tell you that the capacitor is leaking when used at its normal operating voltage. You can tell that it is not shorted; but for very small capacitors it is difficult to tell if the capacitor is open.

You can easily rig a time-constant test for small capacitors, see *Figure 3-18*. The capacitor should charge to about 2/3 the supply voltage (V) in a time (seconds) equal to the resistance (ohms) multiplied by the capacitance (farads).

$$T = R \times C$$

Make the resistance high enough so that you can observe the capacitor charging by the upward swing of the voltmeter pointer. Remember that an open electrolytic capacitor would also result in a high ripple measurement of the supply voltage. Use an oscilloscope for the ripple measurement.

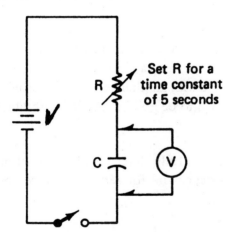

Figure 3-18. Capacitance can be checked with this simple test setup. If the time constant is long, the time taken for the capacitor to charge to about 63% of the supply voltage can be measured accurately. Then C = T/R.

A number of years ago a technician from Florida sent a test that was used for checking leakage of electrolytics. It is passed on to the reader the way he told it, see *Figure 3-19*. It may be useful when it is necessary for you to test electrolytic capacitors. The measured value would vary for different circuits. Note that the voltmeter is being used as a current-measuring device in this application.

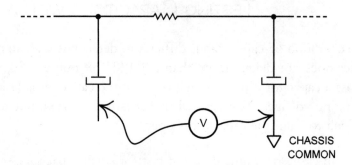

Figure 3-19. A test for checking leakage of electrolytic capacitors.

That test can also be used any time you want to measure electrolytic capacitor leakage currents, and it is a better test than trying to use the ohmmeter 1.5 volts. Technicians often check electrolytic capacitors by bridging them with a capacitor known to be good. The test is illustrated in *Figure 3-20*. The capacitor is bridged with the equipment energized. You should not use this test in a semiconductor circuit.

What you *should* do is make sure the equipment is first turned OFF, connect the capacitor by using convenient alligator clips, then turn the equipment ON. Transients caused by momentary bridging of capacitors when the equipment is ON can destroy semiconductor devices.

We know that some technicians will continue to bridge electrolytic capacitors in energized equipment despite what has been said about it. However, you can increase the safety of that test by turning the equipment OFF before you bridge with another capacitor. Use alligator clips. After you have connected the capacitor known to be good in parallel with one suspected to be bad, *then* energize the system.

The best way to test electrolytic capacitors is to use an ESR meter. An Equivalent Series Resistance meter checks for leakage (parallel) resistance and for series resistance. Some digital VOMs have a built-in ESR tester.

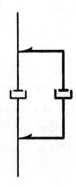

Figure 3-20. Bridging is a popular method of checking electrolytics. However, it is not a good method because it can destroy semiconductor devices in the circuit.

Diode Test with an Ohmmeter

Ohmmeters are sometimes used to check inductive circuits to determine whether or not there is continuity. That test can be misleading. For one thing, it does not show whether there is a short-circuit in the component.

Keep this very important fact in mind: An ohmmeter delivers voltage and current. That current can turn a transistor ON. It can also turn a diode or other active device ON. If there is a high inductance in the circuit, the inductive kickback when removing the ohmmeter can easily be sufficient to destroy a semiconductor device. The best way to test resistors, capacitors and inductors is to use an LCR meter.

Estimating Voltages

Voltmeter measurements are important for evaluating components and circuits, but they are useless if you don't know what the voltage is supposed to be. One way to determine the required voltage is to consult the schematic. Another useful way is to estimate what the voltage should be.

Consider the voltage divider shown in *Figure 3-21*. What voltage should the voltmeter indicate? Even without a schematic, it should be obvious that the amount of voltage across the resistor (R_2) is determined by the proportion of that resistance value to the total resistance in the circuit. Disregarding the very small current from the transistor base, the voltage can be estimated easily as follows:

$$V_B = V_{bb} \left(\frac{R_2}{R_1 + R_2} \right)$$

With a little practice, you can estimate the voltages very quickly.

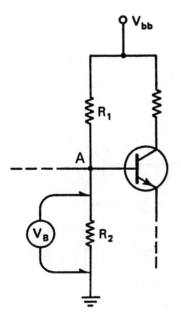

Figure 3-21. You can readily estimate voltages in voltage dividers.

The forward voltage from the emitter to the base of a bipolar transistor is also very important. It is obtained by measuring the emitter voltage and subtracting that value from the base voltage, as shown in *Figure 3-22*. The voltage between the emitter and base of a silicon bipolar transistor in an amplifier circuit is *about* 0.7V. It is lower for a germanium diode and higher for a conducting gallium arsenide PN junction.

Why not simply measure the emitter-base voltage directly? Because you should practice to reduce the amount of time you spend locating circuit troubles by using only one probe.

If you start moving both probes around the circuit you are going to waste valuable time. Remember, the fastest troubleshooting by technicians occurs when they clip the negative lead of the meter to common and make all measurements with the remaining probe. That procedure is especially fast with a DVM that has autoranging.

Amplifier Voltmeter Test

You don't need a schematic drawing to check whether most amplifiers are in working order. There are several basic measurements that are very useful.

In a Class A amplifier the collector (or drain, or plate) voltage should be about one-half the power supply voltage. Therefore, after measuring the power supply voltage (and assuming it is OK) your next voltage measurement is at the collector (or other) output. That will tell you if the amplifier is working and if it is conducting an amount typical of a Class A amplifier.

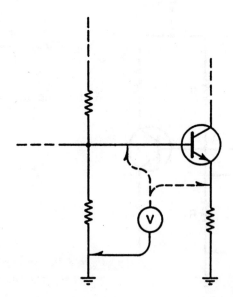

Figure 3-22. This method of measuring emitter-to-base voltage is preferred because the meter is permanently connected to the common point.

If the voltage is too high it means the amplifying device is not conducting enough current. That can be due to excessive bias or a defective transistor. If the voltage is too low it means the transistor is conducting too much current. That can also be a bias problem or a defective transistor.

Once you have located the faulty stage, further tests can be used to determine which component is at fault. The test just given isn't sufficient to warrant the removal of a transistor, but it is a good place to start. You may also determine the condition of an amplifier by measuring amplitude of the AC input and AC output signal with an oscilloscope. Most voltage amplifiers should show a certain amount of gain. That is, the output AC voltage should be more than the input. That is *not* true of power amplifiers, or common-collector (emitter-follower) amplifiers.

You can use an AC voltmeter for that test but you must know the frequency limit of your meter. Usually, a voltmeter can be used at all audio frequencies. Your voltmeter may not be sensitive enough to measure the input signal strength. In that case, a prescaler can be used. That is an amplifier with a specific voltage gain. It amplifies the weak input signal so it can be measured with a voltmeter. As shown in *Figure 3-23*, you have to divide the measurement by the voltage gain of a prescaler.

Note

The term *prescaler* is also used for a dividing circuit that lowers a frequency. It allows a frequency counter to measure a frequency that is above its normal range.

In a power amplifier it is quite common that the amplifier is working very well but the output voltage is not as high as the input voltage. That is typical because the output of a power amplifier should be a signal *current*. Where that current flows through a low value of load resistance, such as in a speaker or transformer primary, the voltage gain may be relatively low even though the power gain is high.

So, when you are making an AC input/output voltmeter test you need to know what kind of an amplifier you are testing. Very often you can tell that simply by the physical size of the transistor. Power transistors are usually large (physically) compared to voltage amplifiers and they operate at a higher temperature. They are often mounted on heat sinks or a metal chassis.

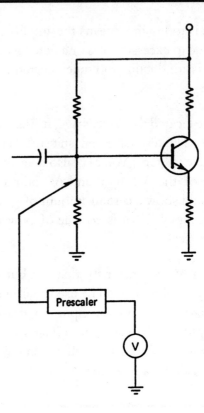

Figure 3-23. Prescalers can be used for measuring unusually low voltages. The prescaler is connected in series with the voltmeter as shown here.

SHORT-CIRCUIT VOLTMETER TEST

The short-circuit voltmeter test is shown in *Figure 3-24*. It is used for bipolar transistor amplifiers. Do not use this test for amplifiers that use FET or tube amplifying devices. It is used by technicians to determine if the bipolar transistor is not only passing current but that it has control over that current.

Think again about measuring the collector voltage. Suppose in a particular case the collector voltage is half the supply voltage, but that is due to the fact that there is an emitter-collector short-circuit (or partial short-circuit) inside the transistor. The transistor is defective even though the collector voltage measurement is correct.

By shorting the emitter to the base, as shown in *Figure 3-24*, the transistor will be turned OFF. The collector voltage should go high. In fact, it should be almost equal to the power supply voltage.

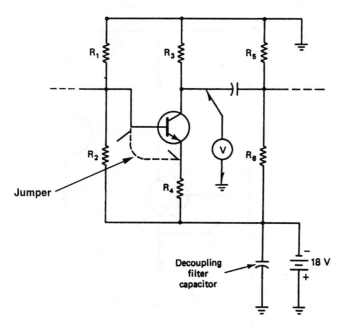

Figure 3-24. The short-circuit voltmeter test.

There are two cases where the short-circuit bipolar transistor test is useless. The first is a Class B amplifier. That type of amplifier is biased at cutoff and will not conduct unless there is an input signal. Since the amplifier is already at cutoff, the short-circuit will not produce any change in the DC output voltage.

The second case where the test is useless (and in fact, it is destructive) is shown in *Figure 3-25*. Observe that the transistors are direct-coupled. The base voltage of the second transistor (Q_2) is equal to the collector voltage of the first. If you connect an emitter-base short-circuit on the first transistor, its collector voltage will rise to a very high value. That high value on the base of Q_2 will surely destroy it.

DIGITAL

Introduction to Combinational Logic

The basic gates are usually used in combinations to make digital circuits. There are two ways these combinations are made: individual gates are wired together to make a circuit; individual gates are combined into a single integrated circuit.

Before discussing truth tables, let's look at another way of remembering the output levels of individual gates.

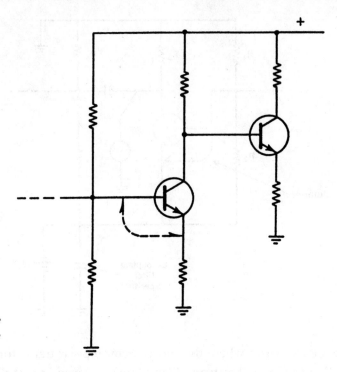

Figure 3-25. In this case the emitter-base short-circuit can destroy the second transistor.

For a two-input AND gate the only way to get a logic 1 output is to have two logic 1 inputs.

For a two-input OR gate the only way to get a logic 0 output is to have two logic 0 inputs.

For an inverter NOT gate the output logic level is always opposite to the input.

For a two-input NAND gate the only way to get a logic 0 output is to have two logic level inputs.

For a two-input NOR gate the only way to get a logic 1 output is to have two logic 0 inputs.

For an exclusive OR gate the only way to get a logic 1 output is to have two different logic level inputs.

For an exclusive NOR gate, also called a *logic-level comparator*, the only way to get a logic level 1 output is to have two identical inputs.

In the following discussion we will use unusual terms to indicate individual outputs. For example, if inputs A and B are combined in an AND configuration we say they are ANDed. As another example, if A and B are combined in a NOR configuration we say they are NORed. Using that non-standard terminology will simplify out discussion of combination. Four examples are given in *Figure 3-26*.

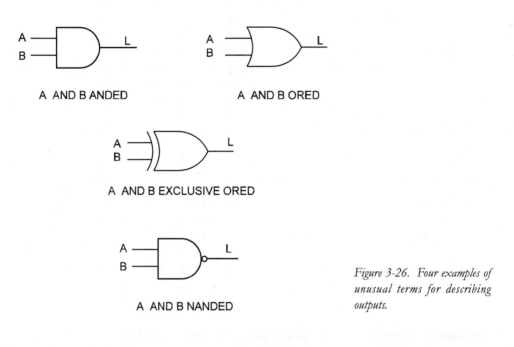

Figure 3-26. Four examples of unusual terms for describing outputs.

Circuits Made of Individual Gates

Let's start with the gates that are combined into circuits. In order to troubleshoot a logic circuit, you should be able to write the output (L) in terms of all possible inputs. You can be sure that all of the possible inputs are supplied to the circuit by using a binary count. You can count in binary numbers using the same technique you use for counting in decimal numbers.

In decimal counting when you have used all of the (Arabic) symbols - 0 through 9 - you have to start new columns. A good example is the odometer in a car that counts in decimal numbers, see *Figure 3-27a*. In binary counting you do the same thing. However, now you only have two symbols: 0 and 1. *Figure 3-27b* shows how a binary count is done. Refer to Appendix B for a longer binary count.

```
   7 6 3 4 1  (Miles)
     ODOMETER
(a)
```

	BINARY COUNT				DECIMAL COUNT
	0	0	0	0	0
	0	0	0	1	1
	0	0	1	0	2
	0	0	1	1	3
	0	1	0	0	4
	0	1	0	1	5
	0	1	1	0	6
	0	1	1	1	7
	1	0	0	0	8
	1	0	0	1	9
	1	0	1	0	10
	1	0	1	1	11
	1	1	0	0	12
	1	1	0	1	13
	1	1	1	0	14
	1	1	1	1	15
1	0	0	0	0	16

Figure 3-27. a) Odometer that counts in decimals. b) Binary count.

(b)

In *Figure 3-28a* there are three inputs to the gate. So, there are $2^3(=8)$ possible inputs. All of the possible inputs are written here: 000, 001, 010, 011, 100, 101, 110 and 111. Using all of the possible inputs you can easily make a truth table for the circuit in *Figure 3-28b*.

Note

Unless you know what signals *should* be present in a system, you cannot be effective in troubleshooting. You troubleshoot by making measurements and comparing your measured values with the values that you know should be present.

Question: Refer to the circuit in *Figure 3-29*. If inputs A and B are always at logic 1, and C is always at logic 0, what is the output of the circuit?

Answer: Logic 1. If A and B are always at logic 1, the AND output is always 1. With a logic 1 input to the OR circuit it doesn't matter if C is at a logic 0 or logic 1. The output of the OR gate will be a logic 1.

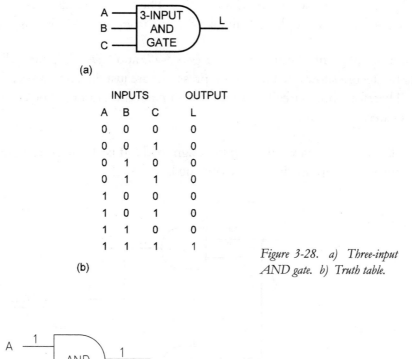

Figure 3-28. a) Three-input AND gate. b) Truth table.

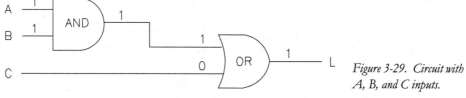

Figure 3-29. Circuit with A, B, and C inputs.

WRITING OUTPUT EQUATIONS FOR COMBINED LOGIC GATES

Start by writing the outputs of each gate. *Figure 3-30* shows how that is done. The output of the AND gate marked X is A AND B (AB). The output of the AND gate, marked Y, is A AND C (AC). So, the two inputs to the AND gate, marked Z, are AB AND AC. The signals are combined in the AND gate to get (A AND B) AND (A AND C). That is written with symbols as ABAC.

Troubleshooting with Meters and Logic Probes 109

Now, we're going to demonstrate how Boolean algebra can be used in logic. This is the only discussion we will have on the subject in this chapter, but it is important enough to give at least one example.

With ordinary algebra you can factor ABAC into ABC. The two expressions are equal, but the second one suggests that a simpler circuit, like the one in *Figure 3-30b*, is possible.

Now this is the important point. The outputs of *Figure 3-30a* and *Figure 3-30b* are ABAC and ABC. Factoring by algebra shows that the two expressions are just different ways of writing the same thing. Therefore, both circuits produce the same output, but the one in *Figure 30b* requires two less gates.

You can see how Boolean algebra would help in design work. It is also useful to the technician because it shows how certain circuits are obtained.

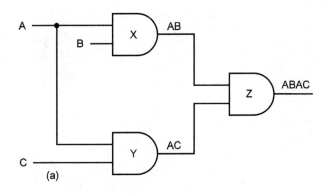

IN BOOLEAN ALGEBRA AA = A
SO, THE OUTPUT CAN BE WRITTEN: ABC
THEREFORE, A 3 - INPUT AND CAN BE USED:

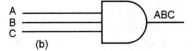

Figure 3-30. a) A basic logic circuit. b) The equivalent gate.

Finding the Output Logic Level of Combined Gates

When you are working in a circuit that is made with combined logic gates, you will need to know the output logic level (1, 0, or pulse) when the inputs are known. You can find the output level by writing the output of each gate, one at a time, until you reach the circuit output. If you have the truth tables for the gates, it is an easy type of problem, see *Figure 3-31*.

Timing Diagrams

Another method of evaluating digital components and circuits is with timing diagrams. They are multitrace signal displays of inputs and outputs of gates and components that are shown on multitrace oscilloscopes.

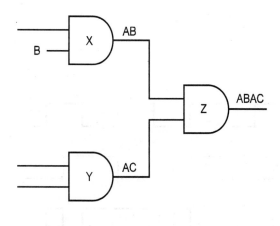

IN BOOLEAN ALGEBRA AA = A
SO, THE OUTPUT CAN BE WRITTEN: ABC
THEREFORE, A 3 - INPUT AND CAN USED:

Figure 3-31. a) Development of the truth table for the circuit in Figure 3-30a. b) The final truth table for the curcuit.

Figure 3-32 shows an example of a timing diagram for an exclusive OR gate. In this example, a square wave is delivered to terminal A and a rectangular wave is delivered to terminal B, resulting in A OR B (the output signal). Remember that the signals for a two-input exclusive OR gate must have opposite polarities in order to get an output of logic 1.

As you can see from the timing diagram, the output amplitude is logic 1 every time the input signals have opposite polarities.

The very basic example in *Figure 3-32* is given to show how timing diagrams work. In practice, those diagrams are far more complicated circuits with as many as eight input and output signals. Normally, the timing diagram for a circuit or system is supplied by the manufacturer. This is *not* a test procedure used by servicing technicians.

EXCLUSIVE OR SYMBOL

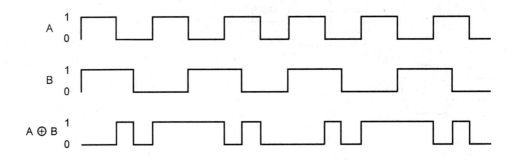

Figure 3-32. The symbol for an exclusive OR gate and an example of timing signals.

Analog Summary

A good technician takes time to check the calibration of the meters used for troubleshooting. It isn't enough just to make measurements for troubleshooting information. The information must be reliable.

Measured information doesn't help unless the *required* value is known. The troubleshooting procedure requires that *measured* values be compared with *required* values.

A taut-band meter movement is rugged and able to maintain accuracy. The restoring force for the pointer is a twisted flat band.

A mirrored scale on an analog meter permits the technician to reduce the problem of parallax.

Some of the measuring techniques used for voltmeters are also useful with oscilloscopes.

Digital Summary

Logic gates are usually used in combinations to make digital circuits. Truth tables are an excellent way to analyze combinational logic circuits.

Truth tables can be written for combinational logic circuits by writing the inputs and output for each logic gate. That must be done with ALL of the inputs and outputs.

Once you have a truth table for a circuit you can determine the output of any combinational logic circuit. Then, you can measure the output for various input logic levels to determine if the circuit is working properly.

A timing diagram is another way of determining if a logic circuit is working properly. It is difficult to produce timing diagrams with the proper input signals unless you have the proper test equipment. Timing diagrams are usually supplied by the circuit manufacturer.

Binary counting is useful for working on digital circuits.

Chapter 3 Quiz

1. Which of the following tests is better for distinguishing between a silicon diode and a germanium diode?

    A. The forward-to-reverse resistance must have a resistance ratio of at least 10 to 1.

    B. The forward voltage across a silicon diode must be about 0.7V and across the germanium diode must be 0.2 volts.

2. The emitter-base short-circuit test should NOT be used with a

   A. high-gain amplifier.

   B. direct-coupled amplifier.

3. Stiction is not a problem with

   A. meters that have jewelled bearings.

   B. taut-band meter movements.

4. With no input signal to an amplifier circuit the voltmeter indicates 6V. With a pure sinewave input signal the voltmeter should display

   A. a higher voltage.

   B. 6V.

   C. a lower voltage.

5. A battery should be tested with a voltmeter

   A. with no load.

   B. with a normal (maximum) load.

6. A technician taps the metal of a screwdriver to the center tap of the volume control in a radio receiver. He is testing the _____ section.

7. A VOM indicates -4 dB at the output of a transmission line and an input of +2 dB. What is the dB loss in the line?

8. Which of the following is less susceptible to extraneous noise?

    A. Digital multimeter

    B. Analog multimeter

9. In order to avoid turning on a semiconductor junction when making a resistance measurement, use the _____ feature.

10. To measure AC current with a volt-ohm-milliammeter so that the voltage scale can be read directly as current, use a resistor having a resistance of _____.

11. What is the approximate input resistance of a VOM on the 100V scale if its meter movement is 50,000 ohms per volt?

12. Current-reading meters can be compared by connecting them

    A. in series in a circuit.

    B. in parallel in a circuit.

13. Demodulator probes are used when the modulated signal being measured

    A. is above the frequency range of the meter.

    B. does not have a 0V average value.

14. What scale should be avoided when connecting an ohmmeter across a PN junction?

15. Which of the following is more rugged?

    A. Taut-band meter movement

    B. Jewelled meter movement

16. Do not use the short-circuit test in what two cases?

17. What type of signal is used for the voltmeter distortion test?

CHAPTER FOUR

Troubleshooting with an Oscilloscope

Overview

An oscilloscope is the most versatile test instrument on the workbench. Here are some of the measurements and displays that an oscilloscope can perform:

Measurements

    Current (DC and AC)

    Voltage (DC and AC)

    Time intervals

    Rise time and decay time of a square wave

    Bandwidth

    Frequency and frequency differences

    Percent modulation

    Phase angles

Displays

>Signals in either a time domain or a frequency domain

>Waveform distortion

>Linear distortion of amplifiers

>Frequency distortion of amplifiers

>Phase distortion of amplifiers

>Frequency response of amplifiers

>Characteristic curves of components

>Alphanumeric characters

>Timing of digital signals

Despite its usefulness in performing a wide variety of service tasks, some technicians prefer to use a voltmeter instead of an oscilloscope. One of the most important disadvantages of the oscilloscope has been, until recently, its physical size. It has not been as portable (or at least as conveniently portable) as a voltmeter. Despite the introduction of new small scopes, they are still expensive compared to VOMs. For troubleshooting in the home, the voltmeter is still favored by some technicians.

An oscilloscope is more difficult to use than a voltmeter. That is no obstacle for a good technician, but many busy technicians cannot take the time to explore its full range of applications. In this chapter some of the techniques for using scopes for servicing are discussed.

OBJECTIVES

>What is a frequency domain display and how is it used?

>What is a time domain display and how is it used?

>When is an alternate mode used with oscilloscopes?

>When is a chop mode used with oscilloscopes?

What is an RS flip-flop?

What is a clock circuit and how is it checked?

What is a JK flip-flop, and what does a truth table tell about it?

What is a Schmitt trigger, and what is it used for?

Why is a delayed sweep necessary?

How can a half-cycle average be determined from a waveform display?

How can an RMS value be determined from a waveform display?

COMPARISON OF OSCILLOSCOPE TYPES

Recurrent sweep scopes are by far the cheapest and easiest to use for waveform display.

A block diagram of a typical recurrent sweep oscilloscope is shown in *Figure 4-1*. In this type, a sawtooth waveform is used to produce a linear horizontal sweep. That sweep is synchronized with the incoming signal by a control called *sync amplitude*, or *sync*. It samples the incoming signal on the vertical input terminals and uses it as the synchronizing signal for the horizontal sweep.

In the normal use of a recurrent sweep oscilloscope, the sync control must be adjusted for the *minimum* amount of sync signal necessary to lock the horizontal sawtooth oscillator onto the waveform being observed. Excessive sync input will cause distortion of the displayed waveform. As shown in the block diagram, it is possible to use an external signal to synchronize the internal sweep. It is also possible to deliver an external sweep signal with the internal sawtooth generator disconnected.

Both the vertical and horizontal amplifiers have two choices of inputs: DC and AC. With the AC input, the signal goes through a capacitor. That prevents any DC offset voltage from getting into the amplifier. In some cases, that DC offset could be sufficient to deflect the beam off the face of the scope.

In the scope of *Figure 4-1*, the sweep generator output passes through the horizontal amplifiers. That way the attenuator and position controls of those amplifiers can be used to control the sweep.

The blanking control shuts the beam off during retrace.

Memory scopes can display a waveform and retain that waveform over a period of time. Also, with some memory scopes the waveform can be printed on a computer printer. Earlier memory scopes employed a special cathode ray tube to retain the display. Today, the signal can be digitized and stored in Random Access Memory (RAM). The numbered parts that represent the waveform can be recalled at any time.

Digital logic scopes are specifically designed to show logic timing diagrams. They are characterized by having a large number of traces, often six or more.

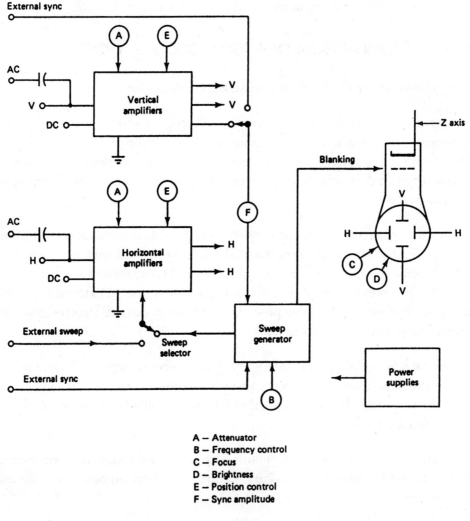

A — Attenuator
B — Frequency control
C — Focus
D — Brightness
E — Position control
F — Sync amplitude

Figure 4-1. Block diagram of a recurrent sweep oscilloscope.

In addition to the scopes just mentioned, some specialized oscilloscopes are readily available. As an example, the phase relationships between signals in a color television receiver are very important for proper display of the color picture. Special oscilloscopes, called *vector oscilloscopes,* can be used to display and measure those phase relationships.

Radar displays are also shown in oscilloscopes used for measuring distances, altitudes and angles.

Triggered sweep scopes require an input signal to start the trace. (Most can be triggered by noise without an external signal.) A block diagram of a typical triggered-sweep scope is shown in *Figure 4-2*. Since the signal starts the trace, the display is very stable. In most triggered-sweep scopes, the time base is calibrated in seconds, milliseconds or microseconds. In the calibrate function, those periods of time are very accurately controlled.

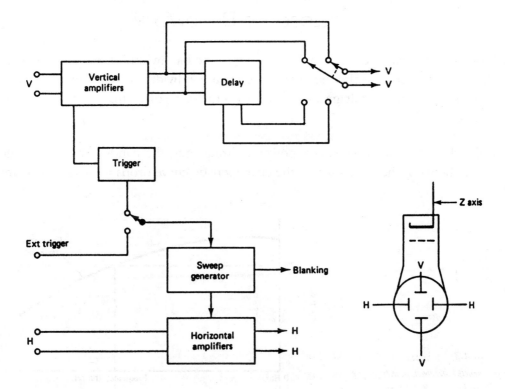

Figure 4-2. *Block diagram of a triggered-sweep oscilloscope.*

Except for the triggering feature, the block diagrams of the recurrent and triggered sweep scopes are very similar. The control for the horizontal sweep in triggered sweep scopes is marked in units of time. Accurately controlled time bases make it easy to measure the frequency. The procedure is to measure the time (T) for one cycle, then calculate the frequency using the equation:

$$f = 1/T$$

*Dual-trace oscilloscope*s can display two time bases at the same time. In most common oscilloscopes that is accomplished by using an internal electronic switch that switches back and forth between the two waveforms being displayed. The switching is so rapid that the eye perceives the traces as being on at the same time. Dual-gun scopes have two separate electron guns and separate deflection plates.

ANALOG

OSCILLOSCOPE BANDWIDTH

Oscilloscopes are advertised and sold according to the bandwidth of frequencies that can pass through their vertical amplifier. Remember that bandwidth is the range of frequencies between points where the voltage amplitude drops to 70.7% of its maximum voltage value, see *Figure 4-3*.

When the characteristic curve shows power vs. frequency, the bandwidth is the range of frequencies between the points where the characteristic curve drops to 50% of maximum.

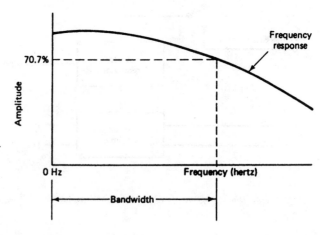

Figure 4-3. Bandwidth is the range of frequencies between points where the voltage drops to 70.7 % of maximum. In this illustration, bandwidth is measured from 0 Hz.

Bandwidth is a very important consideration, especially when observing nonsinusoidal inputs. Consider the case of a square wave signal at a fundamental frequency. As shown in *Figure 4-4*, a square wave consists of a fundamental pure sine wave form and a large number of odd sine wave harmonics. In order to reproduce a reasonable display of the square waveform, it is necessary for the vertical amplifier of the oscilloscope to pass at least eight or ten of the harmonic frequencies. If an amplifier cannot pass those harmonic frequencies, it will distort the square wave form. That is the basis for the qualitative square wave test described in Chapter 1.

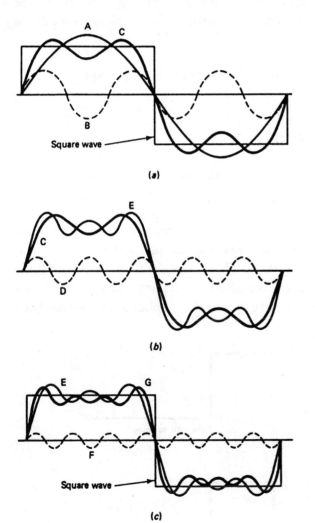

A — Fundamental
B — Third harmonic
C — Fundamental plus third harmonic
D — Fifth harmonic
E — Fundamental plus third and fifth harmonics
F — Seventh harmonic
G — Fundamental plus third, fifth, and seventh harmonics

Figure 4-4. Makeup of a typical square wave. In reality, the harmonics would extend to infinity to make a perfect square wave.

Another reason for needing a wide bandwidth is the reproduction of a glitch. A glitch is an undesired transient voltage having a very short time period. For example, a glitch may have a period of only 10 or 15 nanoseconds. To display the glitch, the vertical amplifier must be able to pass the very short pulse. This means it must pass high harmonic frequencies.

Two limiting factors must be taken into consideration. One is the persistence of the scope screen. Persistence is a measure of how long a trace is visible after the electron beam passes. If the persistence is too long, you won't be able to see rapid changes in the waveform being displayed. The second important limiting factor is the bandwidth of the vertical input amplifier. It must pass all of the harmonics that produce that pulse waveform. If the vertical amplifier cannot pass the range of harmonics associated with the glitch, you will not see it displayed.

A simple circuit that you can construct to determine your scope's ability to display a glitch is discussed in Chapter 10.

DELAY

As shown in *Figure 4-5*, you can't see the rise time of a first cycle when the scope doesn't have a built-in delay.

You could decrease the sweep speed so you can see two cycles. That way you can see the rise time of the first full cycle. However, the rise time is easier to measure with the single pulse display because it takes up a greater part of the trace. That is especially true if you are using the oscilloscope sweep expander. Remember that using the sweep expander to display rise time requires that you must divide the measured rise time by the number of times it has been expanded.

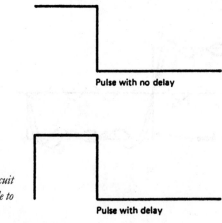

Figure 4-5. Unless there is a time delay circuit in the horizontal amplifier, you will not be able to see the leading edge of a step function.

Frequency vs. Time Domain

Waveforms have three dimensions: amplitude, frequency and time elapsed. Normally, it is only possible to graph two of these dimensions on a flat piece of paper graph or oscilloscope screen. So, the two most commonly used graphs are given as separate displays.

Figure 4-6 shows the relationship between the two-dimensional views as they are seen on an oscilloscope screen. The three-dimensional view in *Figure 4-6a* has only two representative sine wave forms, but any waveform would be located on the graph.

As shown in *Figure 4-6b*, the time axis view has the later times as you look from left to right. That is the time domain display most often seen on scopes.

The *frequency domain display* shown in *Figure 4-6c* is also available on an oscilloscope. Uses of frequency domain displays are discussed in Chapter 5.

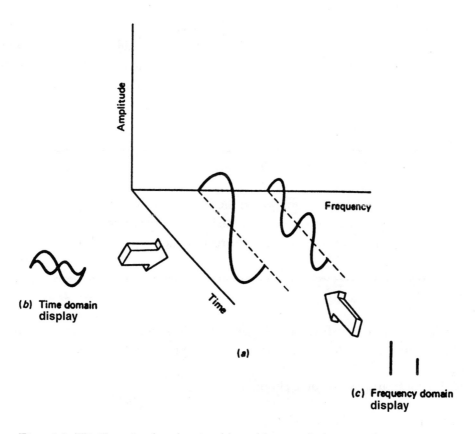

Figure 4-6. This illustration shows how time delay and frequency displays are made.

Oscilloscope Accessories

In addition to the scopes for displaying waveforms, there are usually some provisions for measuring frequency, time and voltage.

Be sure to study the methods of measuring voltage and frequency described in the manual for your triggered-sweep oscilloscope. Those internally calibrated features should be checked periodically to ensure that your equipment is working properly.

Voltage Calibration

To calibrate an oscilloscope, or to verify its internal calibration, use the same techniques descried for voltmeters in Chapter 3. Use the volts-per-division calibration. A 1.5V battery can be used to calibrate the DC vertical input terminal. A filament transformer secondary can be used to check the AC calibration.

A variable AC transformer, sometimes called a variac, is also very useful for checking AC vertical input calibration. On many oscilloscopes there is a calibration voltage-output terminal on the front panel of the instrument. Some technicians prefer to calibrate their instruments with a focused dot, rather than a trace, on the screen. Keep the intensity of that dot as low as possible for best focus.

Caution

If you leave a bright dot on the CRT screen you can get a permanent burn mark in the CRT phosphor coating. *Never* leave the dot on the screen for a long period of time.

Question: The variable transformer in *Figure 4-7* is adjusted for a voltmeter reading of 10V AC. What peak-to-peak value should be indicated on the oscilloscope?

Answer: The peak-to-peak reading should be:

$$2 \times 1.414 \times 10 = 28.28V\ AC$$

Remember that the voltmeter indicates the RMS value. The peak value is 1.414 times the RMS value, and peak-to-peak value is twice the peak value. Actually, it is the square root of 2 times the RMS value. The setup for checking AC voltage calibration shown in *Figure 4-7*.

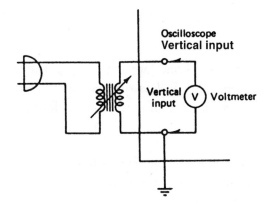

Figure 4-7. Illustration for sample problem.

FREQUENCY CALIBRATION

Question: Assume the oscilloscope sweep is calibrated at 10 milliseconds per horizontal division. How many divisions would one cycle of a 60-hertz sine wave have?

Answer: Using the equation for period (T):

$$T = \frac{1}{f} = \frac{1}{60} = 0.0167 = 16.7 \text{ milliseconds}$$

Therefore, one cycle would take 1.67 divisions.

For an accurate count of milliseconds per division, a frequency counter can be used, see *Figure 4-8*. Once the time base calibration has been checked on one of the divisions, you can assume that the other divisions on the scope time-base adjustments are also accurate.

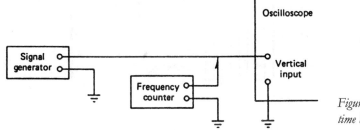

Figure 4-8. Checking the oscilloscope time base for accuracy.

Chop and Alternate

Dual-trace triggered-sweep oscilloscopes have a built-in electronic switch. Its purpose is to switch back and forth between the two displays so rapidly that your eye perceives them to be on at the same time. Electronic switches can also be used with recurrent sweep scopes as an external accessory.

In reality, the electronic switch is nothing more than a square wave generator that switches the trace back and forth between two DC levels. Regardless of whether the electronic switch is built-in or used externally, there are usually two modes of operation: *chop* and *alternate*.

In the alternate mode, one line is traced and then the other line is traced; each trace is displayed alternately. It is used for displaying higher frequencies. At very low frequencies this alternate switching produces a distracting pulsed display.

When this occurs, the *chop* mode should be used. In the chop mode, a portion of one waveform is shown on one trace and then a portion of the second waveform is shown on the other trace. The back-and-forth switching is accomplished very rapidly so that each trace is broken into small parts, see *Figure 4-9*.

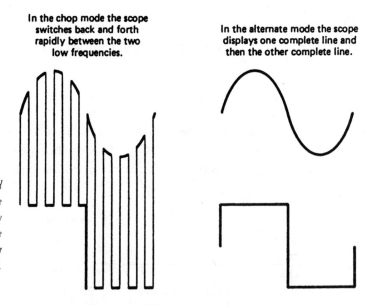

Figure 4-9. Comparison of chop and alternate displays. In the chop mode the trace moves back and forth quickly between parts of the waveform. In the alternate mode one complete waveform is traced and then the other one is traced.

If the two waveforms being displayed have different frequencies, and the frequencies are not exact multiples (like 2xf or 3xf) you may not be able to get a stable display for one of the frequencies on a dual-trace oscilloscope.

Low-Capacity Probes

For most applications, technicians use a low-capacity probe. It puts a capacitor in series with the probe as shown in *Figure 4-10*.

Remember that two series capacitors have a combined capacitance that is lower than either capacitance value. Therefore, the series capacitor combines with the scope capacitance to *reduce* the total capacitance seen by the input signal. For low-frequency applications, such as the low-frequency end of the audio spectrum, a direct probe can be used.

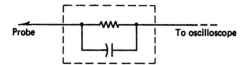

Figure 4-10. Circuit for a low-capacity probe.

A common mistake is to ground the shield of a scope lead coaxial cable at both ends. *Figure 4-11* illustrates the problem. A very small difference in the common voltage between the scope and the system being measured will cause an AC current to flow in the shield. That, in turn, causes an electromagnetic field that injects an undesired interference signal in the scope. That is true for all applications of shielded conductors.

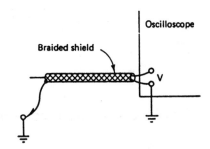

Figure 4-11. Shielded wire should never be grounded at both ends.

Parallax in Oscilloscope Measurements

If you are using an add-on screen graticule for measuring either frequency or voltage with an oscilloscope you must be careful not to introduce a parallax error.

To avoid the problem of parallax, some special-application oscilloscopes like radar scopes, have the graticule marked on the inside of the CRT face. Ordinary oscilloscopes do not have that feature, so it is up to you to avoid the error.

Measuring Current

The procedure for measuring AC current on an oscilloscope is the same as that for measuring voltage. A 1-ohm resistor is placed in series with the current to be measured. The voltage across that resistor is measured with a scope, and the current is numerically equal to the voltage.

If the current being measured is too small to get a sufficient voltage across 1 ohm, 10 ohms or 100 ohms can be used, provided the circuit resistance is no greater than one-tenth the resistance of the circuit being measured.

In order to get the maximum accuracy with scope measurement of a DC or pulse waveform, you should adjust the scope vertical positioning control so that the trace line is at the bottom line of the graticule, see *Figure 4-12*. That gives you the full height of the graticule for making the measurements.

Figure 4-12. Get the maximum accuracy of scope and measurements by making use of the full height available.

Determining Average and RMS Value

Determining Average and RMS Value

You will remember that voltmeters are usually calibrated to display RMS values of sine wave voltages and currents. Some types of meters are designed to measure true RMS values of nonsinusoidal waveforms. An example is the thermocouple meter.

You can measure the average and RMS values of a nonsinusoidal wave with a scope if you are willing to do a few basic calculations. *Figure 4-13* shows how the average value is obtained, and *Figure 4-14* shows how the RMS value is obtained. It is only necessary to take five or ten measurements to get a fairly accurate determination of the average or RMS value. Of course, the greater number of measurements the more accurate the results.

The average value is calculated the same way as finding an average of anything else. Assume the waveform in *Figure 4-13* is for a voltage. Add the voltages, as shown, for one cycle. Then divide by the total number of voltages that you added to get the average value.

The procedure for finding the RMS value is to work backward from the term root mean square:

1. *Square* each value.

2. Find the average, or *mean*.

3. Take the *square root* of the result.

As with the average value, more measurements mean more accuracy. But, you may not need that amount of accuracy for what you are doing.

In most equipment the manufacturer gives peak-to-peak values of voltages or currents when the voltage or current value is needed. However, if you are working in a situation where you need to know the average or RMS value, the techniques shown in *Figure 4-13* and *Figure 4-14* work reasonably well.

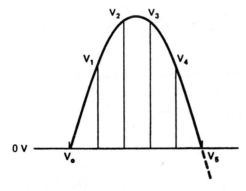

Average value = $\dfrac{V_1 + V_2 + V_3 + V_4 + V_5}{5}$

Note: Do not include V_o.

Figure 4-13. Average value is taken by averaging individual voltage values.

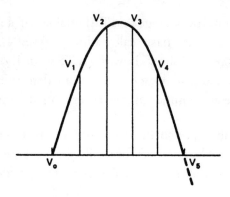

Figure 4-14. *The RMS value is obtained by the method shown here.*

1. Square each value of V.
2. Add the squared values.
3. Take the average of the squared values.
4. Take the square root of that average.

$$\text{RMS value} = \sqrt{\frac{V_1^2 + V_2^2 + V_3^2 + V_4^2 + V_5^2}{5}}.$$

Note: Do not include V_o.

MEASURING/EVALUATING COMPONENTS WITH AN OSCILLOSCOPE

Oscilloscopes are ideal for checking the PN junction of diodes, transistors and other semiconductor devices. They can also be used to determine the operating characteristics of thyristors and other components. The general procedure is to display the current through the device on one axis and the voltage across the device on the other axis. The method of displaying current has been discussed earlier in this chapter. Some oscilloscopes have a *component test* feature. If your scope doesn't have one, it is a matter of making some basic test equipment for doing the job. Test equipment is described in Chapter 11.

AMPLIFIER TROUBLESHOOTING WITH AN OSCILLOSCOPE

There are two types of distortion in amplifiers: frequency distortion and linear distortion. Phase distortion is included in the category of frequency distortion because the amount of phase distortion is dependent on the frequency of operation.

FREQUENCY DISTORTION

Frequency distortion occurs when an amplifier can't pass the range of frequencies it is designed for. The square wave test and the sawtooth test are qualitative. They permit a quick overview of the amplifier's ability to pass its range of frequencies.

The point-by-point test shown in *Figure 4-15a* is accurate but time consuming. The signal generator is tuned over the range of frequencies. The amplitude of the output is plotted for each frequency. The resulting graph is often plotted on semilog paper. It is linear in the vertical direction and logarithmic in the horizontal direction. That permits details over a wide range of frequencies to be explored.

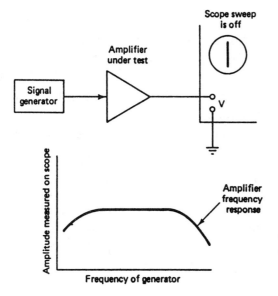

Figure 4-15a. The amplifier frequency response can be obtained by making point-by-point measurements and then graphing the result. Notice that only a vertical line is used for making the voltage measurement.

For each measurement taken, it is necessary to ensure that the signal generator amplitude has not changed when the frequency was changed. That step is not necessary for the more expensive signal generators that have regulated output amplitudes.

For a quick test the graph need not be plotted. Just run the generator frequency control over the range of required frequencies while observing the amplitude displayed on the scope. Turn off the horizontal amplitude control for this test since you are only interested in amplitude.

With the generator connected to the scope amplifier, observe the amplitude displayed on the scope as you range through the frequencies of interest. The length of the line, representing amplitude, should be constant throughout the frequency range.

Then, connect the scope to the output and adjust the generator through the range of frequencies again. Remember, the bandwidth is the range of frequencies where the length of the line does not drop below 70% of maximum. If you see a bright spot on either end (or both ends) of the line, it means the output wave is distorted and the test is not valid. You should reduce the input signal generator amplitude and try again.

The frequency response curve of an amplifier can be plotted directly on an oscilloscope by using the scope in the frequency domain. That procedure is described in Chapter 5.

Instead of manually sweeping the signal generator through the range of frequencies of interest, you can use a sweep generator. It automatically sweeps the range of frequencies introduced to the amplifier. The amplifier output is delivered to the vertical input of the oscilloscope. The generator internal sweep signal must also be connected to the external horizontal input terminal on the scope. This assures that the position on the trace corresponds to the generator frequency at that point, see *Figure 4-15b*.

Using the sweep generator test you will get the same pattern as for the point-by-point test. Function generators often have the sweep feature. However, the output of low-priced function generators is seldom a pure sine wave, and that may distort the sweep display on the scope.

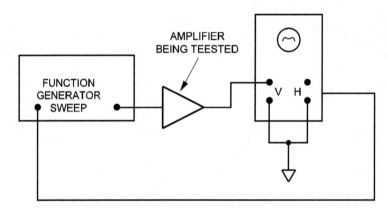

Figure 4-15b. The function generator sweeps through the range of frequencies, and at the same time it sweeps the oscilloscope beam from left to right. The scope display shows the bandwidth (characteristic curve) of the amplifier.

LINEAR DISTORTION

Linear distortion occurs in an amplifier when the output waveform is not an exact representation of the input waveform. The input and output waveforms can be out of phase by 180 degrees, but they must have exactly the same waveform in a low-frequency linear amplifier. Ideally, Class A amplifiers should have no linear distortion. In the real world there is no such thing as an amplifier that is 100% distortion free. However, some designs come very close.

High-frequency linear amplifiers of the type used in transmitters may not be operated Class A. They are designed to produce a very high gain. So a linear amplifier in a transmitter system may actually be a Class B amplifier or in some cases, even a Class C amplifier.

The principle of operation is shown in *Figure 4-16*. The input signal to this Class B amplifier is a pure sine wave. The output is a half-wave rectified sine wave. It does not look anything like the sine wave input. That intended distortion is typical of Class B amplifiers.

The tank (parallel LC) circuit in the output of the Class B amplifier is resonant to the same frequency as its input signal. The half-wave variations produce high-amplitude circulating currents in the tank. That signal is coupled to the transformer secondary as a pure sine wave.

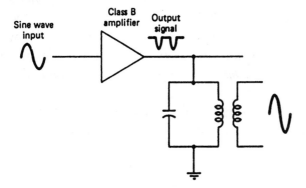

Figure 4-16. This Class B amplifier produces a full sine wave output.

Assume that the amplifier is linear and operating in the low-frequency range. One quick test for linearity uses a lissajous pattern. It is a good idea to review the generation of lissajous patterns before describing this test.

Figure 4-17 shows a test setup to produce a lissajous pattern on an oscilloscope. The scope is set so that the horizontal input signal replaces the internal sweep circuitry. In other words, there is no sawtooth time base on the scope.

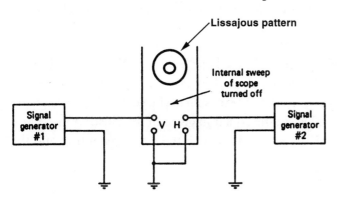

Figure 4-17. The lissajous patterns on an oscilloscope can be obtained using this test setup.

Troubleshooting with an Oscilloscope

The horizontal deflection of the beam is produced by the signal delivered to the horizontal input terminals. On some oscilloscopes that is accomplished by turning the internal sweep off, or to *external*. In other scopes it is called the X-Y operation.

In order to get a true lissajous pattern, it is necessary that the signal at the horizontal plates produce exactly the same amount of deflection as the signal at the vertical plates. You can accomplish that by feeding the two signals to the scope one at a time and adjusting for a specific amount of deflection.

Note: Lissajous tests are *not* valid unless the two inputs are adjusted, one at a time, so they produce the same amount of deflection on the oscilloscope.

The test setup in *Figure 4-17* should produce a perfect circle if the two frequencies are identical and they are 90 degrees out of phase. This provides a test to determine if there is distortion in a sine wave. If you are sure that one of the two inputs is a perfect sine wave, then you can determine if the other is a perfect sine wave (or a distorted sine wave) by using the lissajous test. *If you can't get a perfect circle, the unknown waveform is not a pure sine wave.*

If the frequencies are the same, but the phase angle is other than 90 degrees, the lissajous pattern will be somewhere between an ellipse and a straight line. Some of the variations are shown in *Figure 4-18*.

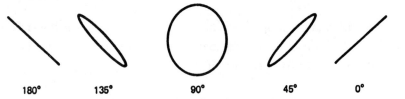

Figure 4-18. Various lissajous patterns obtained when the waveforms are pure sine wave and equal frequenciy. Only the phase between the two waves is different.

Because the out-of-phase sine wave produces a specific pattern, it is possible to determine the phase difference between the two sine waves as a quantitative measurement. The procedure is shown in *Figure 4-19*.

With the two waves producing the ellipse, center the ellipse on the X-Y axis. To do that, count from the center in each direction horizontally and from the center in each direction vertically to make sure that both the X and Y axes are crossing in the center. Then, measure A and the projection of B. Use those values in the equation given in the illustration. If you have a scientific calculator, simply divide A by B. Then push the *inverse sin* button and the phase angle will be displayed.

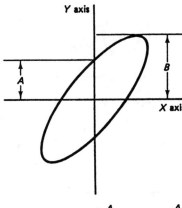

Phase angle = $\phi = \sin^{-1} \frac{A}{B}$, or = arcsin $\frac{A}{B}$.

Figure 4-19. The actual phase difference between two waveforms can be obtained from the lissajous pattern. The procedure is shown here.

You can use the lissajous pattern to calibrate a signal generator frequency against the frequency of one that is known to be accurate. If the input and output frequencies are not the same, the lissajous pattern takes on more complicated displays, as shown in *Figure 4-20*. You can find the ratio of the vertical to the horizontal frequency using the basic technique shown in *Figure 4-21*. Determining frequency ratios by the method shown in *Figure 4-21* is an interesting trick. However, it has very little practical application in troubleshooting.

1. Draw horizontal and vertical lines as shown.
2. $\dfrac{\text{No. of times pattern touches horizontal line}}{\text{No. of times pattern touches vertical line}} = \dfrac{\text{Vertical frequency}}{\text{Horizontal frequency}}$
3. $\dfrac{\text{Vertical frequency}}{\text{Horizontal frequency}} = \dfrac{3}{1}$

The vertical frequency is three times the horizontal frequency.

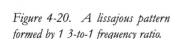

Figure 4-20. A lissajous pattern formed by 1 3-to-1 frequency ratio.

Figure 4-21. The procedure for finding the ratio is shown here.

THE AMPLIFIER LISSAJOUS TEST

Having reviewed the principle of the lissajous pattern, we can now use the pattern to test for amplifier distortion. The setup is shown in *Figure 4-22*. Note that the weaker input signal of the amplifier input goes to the oscilloscope vertical input. In many scopes the vertical input has more gain, so it is better able to amplify that weaker signal.

Troubleshooting with an Oscilloscope

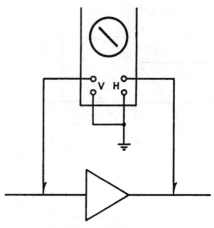

Figure 4-22. This lissajous setup can be used for finding distortion in an amplifier.

Set vertical and horizontal deflections equal. Then set scope horizontal to external input (may be called X-Y input).

Remember that the horizontal and vertical deflections must be equal to get a true lissajous pattern. The pattern of an audio voltage amplifier should be a straight line when the frequency used is about 1000 Hz. At higher frequencies the pattern will open to an ellipse as the generator output frequency is increased. That indicates some phase shift distortion. The specifications for the amplifiers will determine how much phase shift distortion can be tolerated and at what frequency.

The human ear is not sensitive to phase shift distortion, so it does not usually represent a serious problem with audio amplifiers. For other types of amplifiers it may be very important. As shown in *Figure 4-23*, nonlinear distortion will produce a pattern other than a straight line. Remember that bright ends on either or both ends of the pattern indicate clipping distortion.

The lissajous test works best with a single-frequency input. If you are looking at a range of frequencies, such as the output signal from an audio test tape, then the various frequencies will produce various amounts of phase distortion. You won't get a straight line. This can still be a very valuable test.

INTERMODULATION DISTORTION

Intermodulation distortion is another form of nonlinear distortion. It occurs when two different frequencies are applied to an amplifier and those frequencies heterodyne. Heterodyning or modulation can only take place in a nonlinear stage. If it does take place in an amplifier that is supposed to be linear, heterodyning nonlinear distortion is occurring.

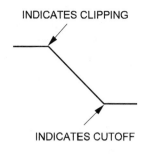

INDICATES CLIPPING

INDICATES CUTOFF

ON THE OSCILLOSCOPE IT MAY LOOK LIKE THIS:

EXAMPLE OF LINEAR DISTORTION

Figure 4-23. Measurement of distortion using lissajous patterns.

Typically, an intermodulation distortion test involves putting two frequencies into the amplifier and then looking at the output on a frequency domain display, see *Figure 4-24*. A special piece of test equipment, called a *spectrum analyzer*, can be used to simplify this test.

Intermodulation distortion is very undesirable because it produces frequencies that are not in the original signal. However, there is some controversy about the test. The question is: Which two frequencies do you use for testing? Different technical groups support different frequencies. So, when you are looking at the specifications for an amplifier, the intermodulation distortion has to be defined in terms of which input frequencies are used.

You might wonder why it is necessary to use tests like the lissajous and intermodulation distortion. Why not just simply apply a sine wave to the input of an amplifier and then see if you get a sine wave at the output? The reason it can't be done is very important. The human eye cannot see distortion in a sine wave as readily as it can see distortion in a circle and on intermodulation patterns. As a matter of fact, the sine wave can be distorted as much as 15% and still look quite acceptable to that human eye.

If you perform the lissajous pattern distortion test using a function generator, you may be quite surprised at the amount of distortion present. You need a known pure sine wave for comparison in this test. The reason is that many function generators produce a *sine wave* by a frequency synthesis method. In that method the sine wave is automatically constructed, point-by-point, but it is not a *pure* sine wave.

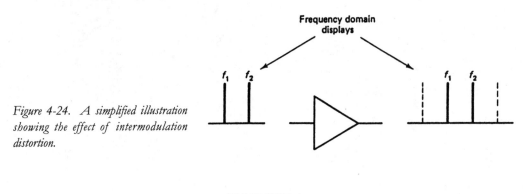

Figure 4-24. A simplified illustration showing the effect of intermodulation distortion.

DIGITAL

THE RS FLIP-FLOP

Figure 4-25 shows two versions of a very popular circuit made by combining basic gates. The circuit is known by the names RS flip-flop, SR flip-flop, latch and bounceless switch. Although it is discussed as a circuit made with basic gates, you can also buy versions of it on a single chip. When a number of gates are mounted on a single chip, the circuit is called an integrated circuit (IC).

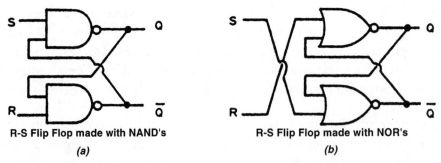

R-S Flip Flop made with NAND's
(a)

R-S Flip Flop made with NOR's
(b)

Figure 4-25. a) an NAND filp-flop b) an NOR flip-flop. Both are examples of RS flip-flops (sometimes called SR flip-flops).

As a general rule, the TTL versions are made with NAND gates and the CMOS versions are made with NOR gates. That affects the method of switching, so the circuits will be compared in this section.

The letters S and R at the input terminals stand for SET and RESET. The letters Q and NOT Q at the output terminals have no special meaning. You will sometimes see 1 and 0 instead of Q and NOT Q. Don't confuse those circuit identifiers with logic levels 1 and 0.

For any flip-flop there are only two possible output states: high and low. A high condition occurs when the Q terminal is at logic level 1 and the NOT Q terminal is at logic level 0. A low condition occurs when the Q terminal is at logic level 0 and the NOT Q terminal is at logic level 1. *Figure 4-26* shows the two conditions that are possible, and it includes a table of names that may be used instead of high and low.

You can analyze the two conditions of the NAND flip-flop by using the same approach that you used to determine the logic level outputs of combined logic circuits. Before starting this study, you should know that the normal logic level of both input terminals is logic 1 for the NAND RS flip-flop. It is switched from one condition to another by switching either S or R to logic 0. The input can be momentarily switched, then returned to logic 1, or it can be permanently switched. In either case, the condition of the flip-flop can only be changed if the proper input terminal is switched to logic 0.

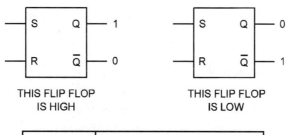

CONDITION	NAME OF CONDITION
Q = LOGIC 1 Q̄ = LOGIC 0	THE FLIP FLOP IS SAID TO BE: ON, OR HIGH, OR IN A LOGIC 1 CONDITION
Q = LOGIC 0 Q̄ = LOGIC 1	THE FLIP FLOP IS SAID TO BE: OFF, OR LOW, OR IN A LOGIC 0 CONDITION

Figure 4-26. Conditions for RS flip-flops in a high and low condition.

Troubleshooting with an Oscilloscope

The normal logic level of the NOR RS flip-flop input terminals is logic 0. An important difference between NAND and NOR flip-flops is that the input terminals are normally at logic 1 for NAND RS flip-flops and logic 0 for NOR RS flip-flops. If a square wave is used to switch the flip-flop, the NAND type will switch on the trailing edge when the square wave goes from 1 to 0. The NOR RS flip-flop will switch on the leading edge when the square wave goes from 0 to 1.

These statements regarding the normal input logic levels are in reference to the basic circuits of *Figure 4-25*. By adding NOT gates to the S and R input terminals the normal input conditions will be reversed. That is likely to be done for the NAND type so that positive logic can be used, see *Figure 4-27*. With positive logic, the flip-flops are switched with logic 1, and they are in their normal condition with logic 0.

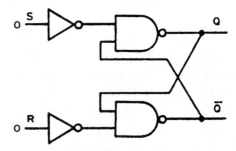

Figure 4-27. Use of NOT gates to change the NAND flip-flop to an active-high condition.

An advantage of positive logic is that when the RS flip-flop is in the resting state, sometimes called the memory state, zero volts is supplied to the input terminals. Therefore, no input power is needed. This can be an important consideration with a memory IC that uses over a million RS flip-flops.

Now let's look at the signals for a NAND SR flip-flop and see how it is switched. *Figure 4-28* shows the NAND flip-flop with logic levels. The numbers are the normal values when the flip-flop is in the high condition. If you switch terminal S to 0, the output of the upper NAND will still be 1 because it will have two 0's at the input. Therefore, there will be no change in the input and output logic levels of either gate when S is switched to 0, provided the flip-flop is in the high condition.

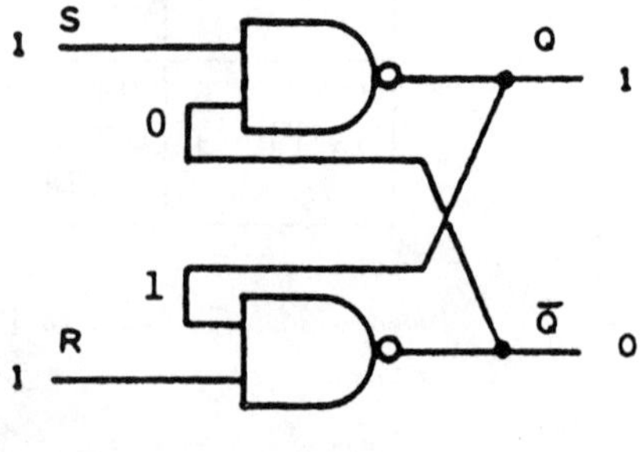

Figure 4-28. Standard conditions for an NAND flip-flop. The input terminals are at the standard logic level and the flip-flop is high. Switching S will not change its condition.

142   Electronic Troubleshooting and Servicing Techniques

Returning the S terminal to its normal logic 1 value does not affect the condition of the flip-flop. To summarize, the S terminal was switched from 1 to 0 and back to 1 on a NAND flip-flop, but there was no change in the flip-flop condition.

Now let's see what happens when you switch lead R to logic 0. If you switch the R terminal to logic 0, as shown in parenthesis in *Figure 4-29*, the output of the lower NAND gate changes to 1. That, in turn, changes the condition of the upper NAND because it will now have two logic 1 inputs and a logic 0 output.

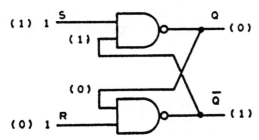

Figure 4-29. *The NAND flip-flop switched to a low condition by changing terminal R to logic 0. Numbers not in parenthesis show condition when R is switched to 0.*

Returning R to 1, as in *Figure 4-30*, does not affect the condition of the NAND flip-flop. Furthermore, if you switch lead R to 0 when the flip-flop is condition low, as shown in

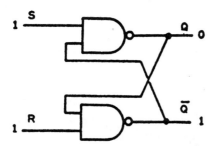

Figure 4-30. *An NAND flip-flop.*

Figure 4-31, it will not change its condition. The only way to get the flip-flop back to the high condition is to switch terminal S to 0. You can verify that by changing the input of the terminal S to 0 in *Figure 4-31* and writing the resulting logic levels.

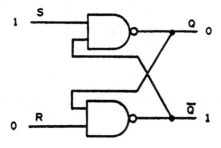

Figure 4-31. *An NAND flip-flop in a high condition. If R is switched to 0 again, it has no effect on the output.*

Troubleshooting with an Oscilloscope

Figure 4-32 shows a summary of input logic levels that will change the condition of an NAND flip-flop. It also shows the changes in input logic levels that have no effect on the condition of the flip-flop. Study that illustration very carefully until you are sure that you understand this action of an RS flip-flop.

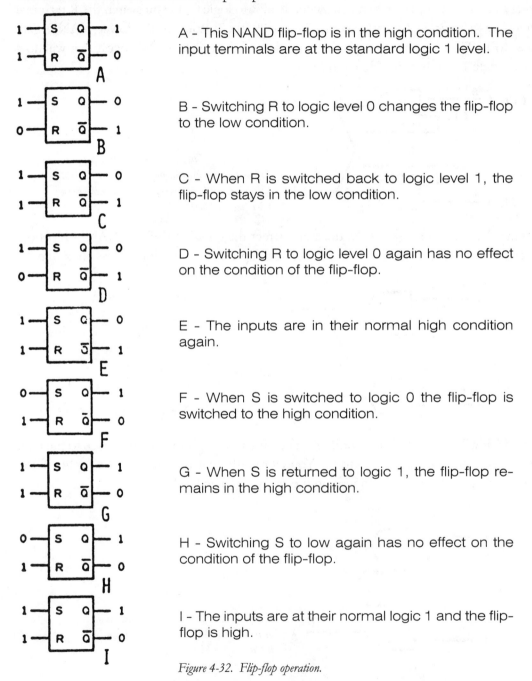

A - This NAND flip-flop is in the high condition. The input terminals are at the standard logic 1 level.

B - Switching R to logic level 0 changes the flip-flop to the low condition.

C - When R is switched back to logic level 1, the flip-flop stays in the low condition.

D - Switching R to logic level 0 again has no effect on the condition of the flip-flop.

E - The inputs are in their normal high condition again.

F - When S is switched to logic 0 the flip-flop is switched to the high condition.

G - When S is returned to logic 1, the flip-flop remains in the high condition.

H - Switching S to low again has no effect on the condition of the flip-flop.

I - The inputs are at their normal logic 1 and the flip-flop is high.

Figure 4-32. Flip-flop operation.

The conditions for the NAND and NOR RS flip-flops can be described with truth tables. The truth tables are shown in *Figure 4-33*. Note that the normal high condition of the NAND type occurs when the inputs are at logic 1 and the normal high for the NOR type occurs when the inputs are at logic 0. They are starting points for the truth tables. From there, they are switched to a low condition and then back to a high condition by changing the input logic levels delivered to S and R.

The question marks indicate that an important thing occurs when the S and R terminals are both switched to level 0 in the NAND SR flip-flop. The same thing happens when the S and R terminals are both switched to level 1 in the NOR flip-flop. In both cases, the condition of the flip-flop *cannot* be determined. That is also called a *not allowed* condition. SR flip-flops can only be used in circuits where there is no possibility of a *not allowed* condition. If that is a disadvantage in a particular application, the designer will choose a different type of flip-flop.

NAND S-R FLIP FLOP					NOR S-R FLIP FLOP			
S	R	Q	Q̄		S	R	Q	Q̄
1	1	1	0	← NORMAL HIGH →	0	0	1	0
1	0	0	1	← FLIP FLOP LOW →	0	1	0	1
0	1	1	0	← FLIP FLOP HIGH →	1	0	1	0
0	0	?	?	← NOT ALLOWED →	1	1	?	?

Figure 4-33. Flip-flop truth tables for NAND and NOR versions.

CLOCK CIRCUITS

In microprocessors, computers and control systems it is a common practice to use a square wave signal (or a rectangular signal) as a reference for all internal circuits. The square wave is called a clock signal, and it often comes from a multivibrator circuit that is labeled *clock*, or *clk*.

Remember that the NOT gate is actually an amplifier operated at either cutoff or saturation. In a multivibrator circuit made with two transistors, one is in saturation and the other is in cutoff. On each half-cycle, the operations of the transistors reverse.

It seems logical that a NOT gate can be used as a multivibrator, and, in fact, you will find such circuits used for a clock. *Figure 4-34a* shows a very popular example. If a highly accurate clock frequency is needed, as in the case of clocks (for telling time) and stop watches, a crystal may be used as shown in *Figure 4-34b*.

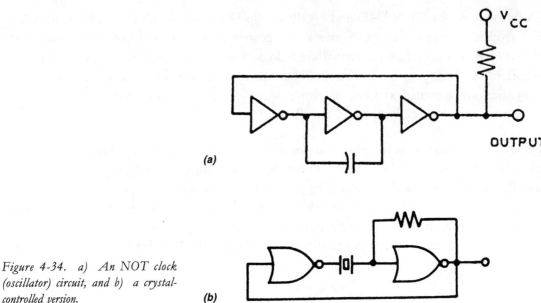

Figure 4-34. a) An NOT clock (oscillator) circuit, and b) a crystal-controlled version.

When you are troubleshooting logic circuitry it is a good idea to check the clock signal. A logic probe can be used to show that a pulse is present, but an oscilloscope may be needed to study waveforms in some troubleshooting problems. A frequency counter can be used to check the clock frequency.

COMBINED CIRCUITS ON AN INTEGRATED CIRCUIT CHIP

There are a great number of circuits made of gates and packaged in integrated circuit form. It would not be possible to list them all in a single chapter. When you get a book of IC logic pinouts it would be a good idea to look through it to see what is available.

An example of an IC package is shown in *Figure 8-8*. The quad NANDs in that illustration are contained in TTL package #7401. As another example, the SN74279 is a TTL package of four RS flip-flops.

JK Flip-Flops

The RS flip-flop has a *not allowed* condition which could make it useless in some applications. The JK flip-flop has no such disallowed input. It is made by taking basic RS flip-flops and combining them with some other basic gates. The overall result is a more elaborate, but also more useful circuit.

Figure 4-36 shows a typical JK flip-flop. Normally, two or more such flip-flops are located in a single integrated package. If you replace the S and R with J and K in the flip-flops of *Figure 4-32*, the operation would be the same. However, there are additional terminals, as shown in *Figure 4-36*, that also affect the operation. There are JK flip-flops in all logic families. The discussion here is for a JK flip-flop in the TTL family.

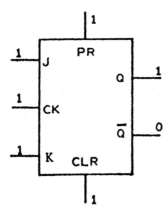

Figure 4-36. *Symbol for a JK flip-flop. Normal voltages shown for the flip-flop in the high condition.*

The CK terminal is used for a clock input signal. The JK flip-flop cannot change by switching J or K to logic 0 unless the CK terminal is switched to 0 at the same time. *Figure 4-37* shows the conditions. When K is switched to 0, but CK is held at 1, the flip-flop condition does not change. However, when the CK terminal is switched to 0 and K is switched to 0, then the condition of the flip-flop changes to low. That arrangement assures that the flip-flop is not accidentally switched with a noise signal, or with a glitch.

The 5th and 6th rows of the truth table in *Figure 4-37* show that the condition of the flip-flop changes when the CK terminal goes to 0. That is indicated by the symbol **. The SET and PRESET terminals are held at logic 1 for this operation.

The PR (preset) terminal is used to switch the flip-flop to high. That is shown in the first row of the truth table. That PR signal makes it possible to switch all flip-flops in a system to high at the same time.

The CLR (clear) terminal is used to switch the flip-flop to a low condition regardless of what condition it is in. The second row of the truth table shows how it works. The reason for having CLR operation is that there are times when you want all of the flip-flops in a system to switch to low at the same time.

The fourth row shows that when both inputs are switched to LOW the flip-flop does not change its condition. Remember that is a *not allowed* condition for the RS flip-flop.

The toggle condition, which is the 7th row of the truth table, shows that if all terminals except the clock input are held at logic 1, then the flip-flop will change condition every time the CLK terminal is switched from 1 to 0. The CLK terminal is then called a toggle.

PR Preset	CLR clear	Clock	J	K	Outputs Q	$\bar{Q}$
L	H	X	X	X	H	L
H	L	X	X	X	L	H
L	L	X	X	X	H*	H*
H	H	**	L	L	Q_0	$\bar{Q}_0$
H	H	**	H	L	H	L
H	H	**	L	H	L	H
H	H	**	H	H	Toggle	
H	H	H	X	X	Q_0	$\bar{Q}_0$

H = High level (steady state)
L = Low level (steady state)
X = Irrelevant
** = Transition from high to low level
Q_0 = The level of Q before the indicated input conditions were established
toggle = Each output changes to the complement of its previous level on each active transition (pulse) of the clock
* = This configuration is nonstable; that is, it will not persist when preset and clear inputs return to their inactive (high) state

Figure 4-37. *Conditions for the NAND flip-flop operation.*

Schmitt Trigger

Figure 4-38 shows the logic symbol for a Schmitt trigger. You probably studied the tube or transistor equivalent at some time. Basically, the Schmitt trigger is used to produce a square wave output for almost any input waveform. A sine wave input will produce a square wave output. The TTL 74132 integrated circuit contains four Schmitt triggers.

Figure 4-38. *Symbol for a Schmitt trigger.*

Analog Summary

Oscilloscopes are somewhat harder to use than voltmeters, but they can be used for a wider range of tests. Qualitative tests require you to make a judgment about the scope pattern you are observing. Quantitative tests provide you with exact values. Both qualitative and quantitative tests are obtainable with oscilloscopes.

Scopes are sometimes used to determine the bandwidth of an amplifier. You can get an approximate bandwidth by measuring the rise time of a square wave into and out of an amplifier. A point-by-point plot is more useful and more accurate, but it takes longer.

Lissajous patterns can be used to determine if there is a pure sine wave. It can also be used to determine frequency ratios, but that technique has very little practical use.

The lissajous test can be used to determine if an amplifier is introducing too much linear distortion.

Harmonic distortion is more difficult to test in amplifiers. The test works better with a spectrum analyzer. The frequencies used for a harmonic distortion test have not been standardized. Harmonic distortion is very important, but you must exercise great care in performing the test.

Digital Summary

Pay close attention to the signals that switch RS and JK flip-flops. Remember that there is a not allowed condition for the RS flip-flop, but not for the JK flip-flop.

Truth tables for flip-flops are very important.

The Schmitt trigger is used to obtain square waveforms from other types of waveforms.

Digital voltmeters are more convenient in many ways than analog meters but having both types is preferable.

CHAPTER 4 QUIZ

1. To measure AC current with a voltmeter or oscilloscope,

    A. connect a 1-ohm resistor in series with the circuit. Measure the voltage across that resistor.

    B. measure the open-circuit across a break in the line and divide by ten megohms.

2. Two types of oscilloscope displays are time domain and _____ _____.

3. When looking at low-frequency waveforms on a dual-trace oscilloscope use the

    A. alternate mode.

    B. chop mode.

4. Which of the following components is connected internally on an oscilloscope AC input?

    A. Capacitor

    B. Inductor

5. What is a blanking circuit used for in an oscilloscope?

   A. Turn off the beam during retrace

   B. Turn off the beam during trace

6. In order to show the leading edge of a square wave of the first cycle of a square wave some scopes have a _____ signal provision.

7. In order to display a glitch your oscilloscope must have a

   A. wide bandwidth.

   B. high vertical amplifier gain.

8. The horizontal trace of an oscilloscope is calibrated for 5 milliseconds per main division. How many main divisions are required for a 100-hertz sine wave?

9. When there is no DC offset the full-cycle average of a 17-volt peak-to-peak sine wave voltage is _____ volts.

10. The bandwidth of a voltage characteristic curve is the range of frequencies between the points that are

    A. 50% of the peak value.

    B. 70.7% of the peak value.

11. The emitter-base short-circuit test should *not* be used with

    A. a high-gain amplifier.

    B. a direct-coupled amplifier.

12. Stiction is not a problem with

    A. meters that have jewelled bearings.

    B. taut-band meter movements.

13. Which type of flip-flop has a *not allowed* condition?

14. What happens when two lows are delivered to the input of a NAND JK flip-flop?

15. Which type of flip-flop can be toggled?

16. Using a X10 sweep expander the rise time of a certain square wave is measured and found to be ten microseconds. What is the true rise time?

    A. 100 microseconds.

    B. One microsecond.

The laws of Boolean algebra are given in Appendix C.

CHAPTER FIVE

Signal Injection and Signal Tracing

Overview

The methods of signal injection and signal tracing are especially useful for troubleshooting systems that have no output. A dead receiver is the example used in this chapter. However, the methods can be used in all electronic systems.

In addition to isolating trouble, it is sometimes necessary to evaluate the system using various tests and measurements. For example, the system may be able to pass a signal from input to output but cannot pass the required range of frequencies. Those necessary tests and measurements are also discussed in this chapter.

Objectives

What are some methods of troubleshooting systems?

What types of test equipment are useful for system troubleshooting?

How is a frequency domain display used for system troubleshooting?

What are some preliminary troubleshooting tests that can be done without test equipment?

What are the precautions to be taken with regard to AFC and AGC/AVC?

SYSTEM TROUBLESHOOTING

This is opinion: *The best troubleshooting aid is the technician's knowledge of how a system (or circuit or component) works.* If you know how something works, you can better understand what is happening when it isn't working.

Technicians sometimes take issue with this viewpoint. Their argument is that you cannot make any money in troubleshooting by analyzing a system. Some technicians even claim that students just out of school can't troubleshoot because they are too heavy in theory and too light in practical methods of finding and repairing troubles.

Perhaps the real truth about troubleshooting is somewhere between the two extremes. Quicker techniques can be used for troubleshooting in most cases, but for those few tough dogs that do not yield to fast troubleshooting methods, the best attack is an analysis based on the technician's knowledge of the system.

In this book, a system is considered to be a combination of amplifiers, oscillators and other circuits. Troubleshooting a dead system requires the isolation of the faulty circuit. After that, the defective component must be located. There are several different ways to locate a faulty circuit.

CUTTING THE TIME REQUIRED FOR SIGNAL INJECTION OR SIGNAL TRACING

System troubleshooting is sometimes done by starting at one end and working toward the other. At least, that is the procedure often described in books. Also, a system diagnostic (discussed in Chapter 6) usually starts at the input end of the system. You can likely save time by starting in the *middle* of the system. An example will be given using the radio, see *Figure 2-24*.

If you inject the signal at the center tap of the volume control and get an output signal at the speaker, you have eliminated the possibility of trouble in all of the stages between the volume control and speaker. In other words, you have divided the receiver in half and reduced the time for troubleshooting.

For a quick injection method, tap the center tap of the volume control with a screwdriver and listen for a clicking noise in the output. Be sure the volume control has been set for high volume when you perform this test. *Figure 5-1* shows this crude method of injecting a signal.

You cannot determine anything about the alignment by using a pulse or the screwdriver test. Some technicians use the blade of a screwdriver for signal injection when they are working away from their bench instruments. To demonstrate how wide a range of frequencies this can produce, try tapping the blade of a screwdriver onto the antenna terminals of a television receiver and note that flashes appear on the screen. You will also hear clicking and static noises in the speaker. Using a screwdriver blade takes a certain amount of skill and experience so that you know what to expect from the particular system you are servicing.

This is not a discussion about how to use screwdrivers instead of test equipment. It is merely included as a time saver. Remember that the screwdriver injection method does not tell you anything about quality of the sound being passed through, or the amount of gain available from each stage.

In both the signal injection and tracing method there should be a change in the amount of gain from the input of the amplifier to the output of the amplifier. Technicians train themselves to listen for this. If they see no change in the gain, further investigation of the stage is recommended. Keep in mind that it may be normal for no stage gain to be noted. An example would be in the case of an emitter-follower stage.

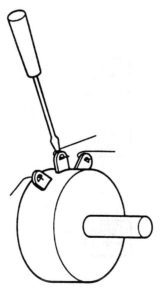

Figure 5-1. A quick injection method for troubleshooting.

Instead of using a screwdriver, you may prefer to use a signal injector. This simple device injects a pulse or square wave into the system being tested. Because of their broad harmonic content, signal injectors can be used in RF, IF and audio stages. Signal injectors can be purchased in kit form, or you can build one from scratch using the circuit described in Chapter 11.

After using the injection method to determine which half of the system is at fault, proceed with either the signal injection or signal tracing procedure.

USING A RADIO AS A SIGNAL INJECTOR

Not all of the circuits in a system are amplifiers. In a radio, for example, there is a local oscillator and a detector. In other systems there may be other nonamplifier circuits.

Since most of the techniques in this chapter are for amplifiers in systems, the methods of locating oscillator and detector faults will be treated separately. Remember, these techniques are applicable to a variety of other systems.

Suppose the oscillator in the receiver of *Figure 5-2* is defective. One method to determine if it is working is to inject a signal generator signal at the proper frequency into the system.

In an AM receiver, like the one used as an example in this book, the oscillator frequency is always 455 kilohertz above the frequency being tuned. So, when they are subtracted in the converter (or mixer), the difference is the 455 kHz IF frequency.

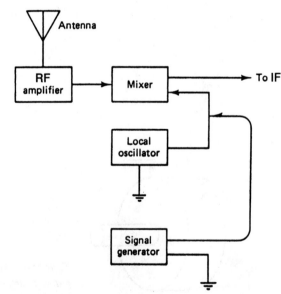

Figure 5-2. A block diagram of the receiver tuner.

Question: A dead AM receiver with a defective oscillator circuit is tuned to a station at 1000 kHz. What oscillator frequency should be injected to make the receiver play?

Answer: Station frequency = 1000 kHz

+ 455 kHz

Oscillator frequency = 1455 kHz

This local oscillator signal should be injected by the signal generator into the converter or mixer stage. The generator signal is injected at the same point as the output of the oscillator circuit. A capacitor is advised to isolate the generator from the receiver DC circuits.

> **Note**
>
> Most signal generators have a built-in isolation capacitor. Check to see if it is present in your signal generator. If so, do not add another isolation capacitor.

Figure 5-3 shows the point of oscillator injection for the receiver used as an example in this book. A quick way to check the oscillator is to position another receiver back-to-back with receiver A, which is being serviced. Receiver B is the good receiver in the simplified test setup shown in *Figure 5-4*. The volume control of receiver A is turned up, and it is tuned to a strong local station. The volume control of receiver B is turned down all the way and it is set to the same station. Under this condition, the local oscillator of receiver B will inject a radiated signal into receiver A. In most cases, receiver A will begin to play. You may have to move the tuning adjustment of receiver B back and forth to use this trick. It is a simple, quick way to determine if the oscillator is working in the receiver being serviced.

SIGNAL INJECTION

Signal injection is used to locate a defective amplifier in a dead system. The idea behind this method of testing, shown in *Figure 5-5*, is to inject a signal into each amplifier one at a time, and observe the output of the system.

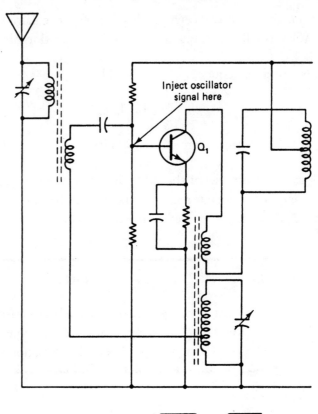

Figure 5-3. Injection of the oscillator signal directly into the circuit.

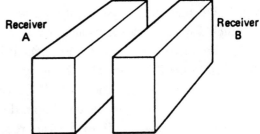

Figure 5-4. One method of checking the oscillator of a receiver is by using the oscillator in a receiver that is known to be working.

If you start by injecting the signal at 1, then 2, then 3, and so forth, you will not get an output signal until you have gone past the trouble. So, if you inject the signal at 2 and do not get an output, but inject it at 3 and do get an output, the trouble must be before 3.

There is an advantage of injecting the signal in the sequence just described. Each time you move the signal source the output signal will get weaker unless you increase the strength of the signal from the signal generator. This is better than starting at the output and moving back. Moving the other way requires a relatively strong signal to be injected at the output. When moving back to the previous amplifier, you can overdrive it enough to cause damage.

What if the amplifiers in the illustration are high-frequency tuned types? In that case, use a signal generator that is adjusted for the frequency that is supposed to pass through the amplifier stages. This will eliminate the possibility, although rare, that the trouble is an improper alignment of the tuned circuit between the amplifiers.

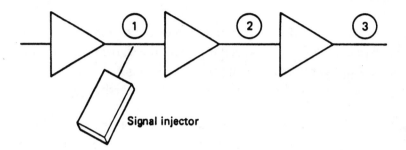

Figure 5-5. Troubleshooting by signal injection.

SIGNAL TRACING

Instead of injecting the signal at each stage and monitoring the output, another technique is to inject a signal at the input and trace the path of the signal through the system. A good oscilloscope can be used to look at the signal at various points along the signal path. This is shown in *Figure 5-6*.

Instead of an oscilloscope, a speaker or headset can be used to trace the signal through an audio system.

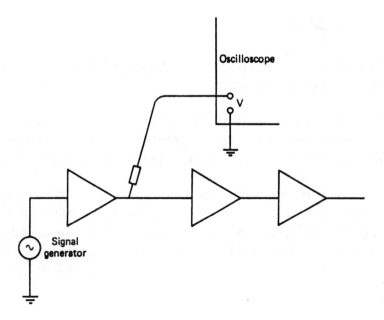

Figure 5-6. Troubleshooting by signal tracing.

Special signal tracing equipment has been manufactured. This will be described in terms of a radio system, but the idea works with any kind of system. As shown in *Figure 5-7*, the signal tracer actually consists of all of the stages that are available in a radio. So, you probe various parts of the receiver with a switch delivering the signal to the corresponding part of the receiver that is in the signal tracer.

For example, suppose you are looking at the output of the detector that goes to an audio voltage amplifier. Your signal tracer would be switched to deliver the signal from the output of the detector to the voltage amplifier of the signal tracer. If you get a signal there, you know that everything from the antenna to the detector is working properly.

In some systems this type of signal tracing equipment becomes very sophisticated and sometimes expensive. This is true for industrial electronic test stations where the equipment is specially designed and put in rack mounts. In some applications this technique is preferred over signal injection.

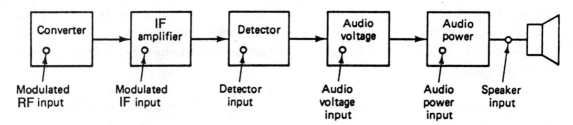

Figure 5-7. The signal tracer is really a special form of radio.

SIGNAL INJECTION WITH THE AMPLIFIER SIGNAL

Return again to *Figure 5-5*. If you have reason to suspect one of the amplifiers, you can use a quick check to determine if that amplifier is defective.

For the purpose of this discussion we will assume that it was the amplifier between 2 and 3. A quick check to see if the amplifier is inoperative is to use a capacitor to connect the signal out of the first amplifier to the input of the third amplifier. This is shown in *Figure 5-8*.

There will be some loss in the gain of the system because you won't have the advantage of the second amplifier. But it is quite often an easy matter to determine if an amplifier is not working by using this procedure. Naturally, you can't use a capacitor that won't pass a signal. For example, if this is an audio amplifier, you should use a 0.2 microfarad capacitor that has a relatively low reactance to the amplified audio signal. Otherwise, the capacitor will not be able to inject the signal.

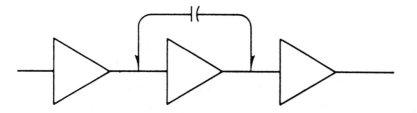

Figure 5-8. A quick check to see if an amplifier is defective.

USING THE PROPER PROBE

If you are following an RF signal, or looking for an injected signal in the RF range, the oscilloscope may not be able to display it. Likewise, if you are looking for transient voltages, such as glitches, the scope may not be able to see them.

In the case of tests for analog amplifiers, where you are using a signal generator with a frequency too high for your scope, you can amplitude modulate the signal. This is done with the internal modulation feature on the signal generator. If it is a function generator, you can add the audio modulation signal from an external point, then use a detector probe on the oscilloscope to demodulate the test signal. The oscilloscope will display the modulation signal, indicating that the RF signal is passing through the amplifier. This works even though you can't see the RF portion of the signal on the scope.

USING THE FREQUENCY DOMAIN DISPLAY

You can get a good idea of the frequency response of an amplifier using the test setup shown in *Figure 5-9*. There is an important restriction on this test. The output of the signal generator must be a constant value over the range of frequencies being tested.

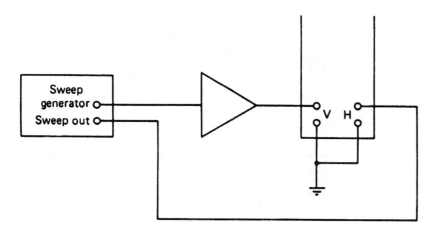

Figure 5-9. The variable DC voltage changes the frequency of the generator. At the same time it changes the position of the trace on the oscilloscope.

It may be necessary to monitor continuously the input amplitude to the amplifier. That will ensure that any changes in the output are due only to the amplifier and *not* to the changes in the signal generator output.

The test procedure is quite simple. The output of the amplifier is delivered to the vertical input to the scope. Then the signal generator is tuned through the range of amplifier frequencies. The amplifier output is monitored. For this test it is a good idea to turn off the horizontal sweep of the scope because you are only looking at the amplitude.

As you adjust the generator through the range of frequencies, you are making sure there is no radical change in the output amplitude. Such a change would indicate an uneven frequency response. Remember that the bandwidth of the amplifier is the range of frequencies between seven-tenths of the maximum voltage amplitude.

It is absolutely necessary that the amplifier does not clip the waveform at any part of the test. This would be indicated by a bright spot on the end vertical line. If that happens, you have to reduce the input and rework the test. Remember, you are looking for undistorted output in the frequency range.

USING THE SWEEP TECHNIQUE

Instead of using a DC voltage for getting the frequency domain display on the oscilloscope, as in the test just described, you can use a sweep generator or function generator with a VCO input. This test setup is shown in *Figure 5-10*. The sine wave voltage causes the function generator, or sweep generator, to sweep back and forth throughout the range of frequencies that you are interested in.

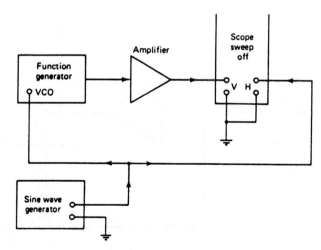

Figure 5-10. Frequency response of an amplifier can be determined with a sweep generator. The oscilloscope must be in the frequency domain mode.

The sine wave generator moves the oscilloscope sweep back and forth at the same rate that it moves the frequency of the generator up and down. So, when the sine wave is at its lowest amplitude, the generator is delivering its lowest frequency. Conversely, at the highest amplitude of the sine wave, the generator is delivering its highest frequency.

One of the problems with the test setup so far is that you don't know just exactly what frequencies you are looking at. You can mark the graticule division for corresponding frequencies throughout the range by starting with the test setup shown in the previous section. Using a DC voltage and a frequency counter, you can mark important points on the graticule corresponding to significant values of frequency.

As an example, suppose you are checking an audio amplifier to determine that it has the ability to pass all frequencies from zero to 100,000 Hz. Using the DC power supply you can determine what voltage is necessary to produce 100,000 Hz output. That would be the maximum positive *peak* voltage of the sine wave injection signal. Likewise, the lowest value would be the one corresponding to the function generator's lowest frequency (usually very near zero hertz).

Having marked this range of frequencies, you can also mark the center frequency. *Figure 5-11* shows a marked graticule.

Now a sine wave is injected with the peak-to-peak signal corresponding to the range of DC voltages that you have just determined. The signal generator will sweep all of the frequencies when that sine wave voltage is applied to the VCO input.

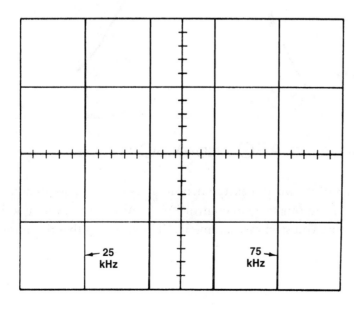

Figure 5-11. You can mark the graticule for specific frequencies on the frequency domain display.

Signal Injection and Signal Tracing   163

A sine wave is not linear, so the frequencies at the high and low ends may tend to be crowded together. You can eliminate this crowding by using a sawtooth voltage for sweeping the frequency, but it isn't usually necessary to get the type of accuracy required for this setup.

When it is necessary, as when aligning an IF stage, a sweep generator may have provisions for marker inputs. These markers are simply fixed frequencies that produce birdies on the trace, see *Figure 5-12*. These birdies tell you exactly where the frequencies of interest are located. The manufacturer will tell you where to set these frequencies so that you can get the desired frequency response.

It requires a high-voltage injected signal, but the marker signals can be introduced into the Z-axis of the scope. The Z-axis is usually located on the back of the scope. It is also called intensity modulation. An input signal to the Z-axis will cause a bright spot on the trace, so it is possible to use this terminal for introducing a marker.

In many cases the Z-axis requires a relatively high voltage, as much as 35V. However, this method of marking has the advantage that it does not distort the frequency domain trace.

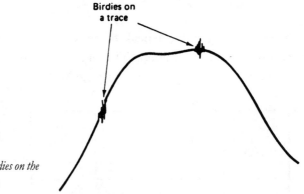

Figure 5-12. Marker birdies on the frequency domain display.

THE ROLE OF AFC AND AGC

If you are going to perform *any* kind of sweep measurement in a system that has an automatic frequency control (AFC) or an automatic gain control (AGC/AVC), it is a very good idea to deactivate these circuits before you start measurements. The reason is that the automatic gain control tends to smooth out any changes in amplitude that would occur as a result of poor amplifier frequency response. Likewise, the automatic frequency control will cause the system under test to track the sweeping frequency. That, in turn, will make it impossible to tell where you are on the frequency domain display.

Deactivating these circuits is usually accomplished easily by the use of a DC power supply. A DC voltage is used to replace the DC output in the closed loop of either circuit. You may be able simply to clamp the AGC voltage by connecting a DC source across it. In other cases it will be necessary to disconnect the AGC/AVC output. In any event it is worth the time that it takes to deactivate these circuits so that you get a better presentation of the amplifier's response.

Noise Generators

A sawtooth and a square wave contain a wide range of harmonic frequencies. This accounts for their usefulness in testing certain types of circuits.

Another signal that has a very wide range of frequencies is *white noise*. White noise gets its name because it supposedly contains all audio and RF frequencies, as in the case of white light which has all visible frequencies. In reality, this is not quite true, but the tradition persists. Pink noise is a noise that has more lower frequencies than higher frequencies. One of the best sources of white noise is a semiconductor diode. Resistors also produce a considerable amount of white noise in a circuit.

In the formal method of aligning a discriminator or ratio detector (two kinds of detector circuits for FM signals), a sweep generator is used to sweep the range of frequencies and the output is displayed on an oscilloscope.

This procedure requires the same test setup as used for viewing the bandwidth of an amplifier on a scope. A quick way to check the amplifier and detector circuits without resorting to the sweep method is to inject white noise from a white noise generator into the circuit, see *Figure 5-13*. With the noise injected, the circuit is adjusted for maximum output noise. When the noise is maximum, the range of frequencies being passed is also maximum.

If you are going to develop this technique, it is a good idea to compare the white noise input with the formal sweep method because it is a qualitative test and requires some experience.

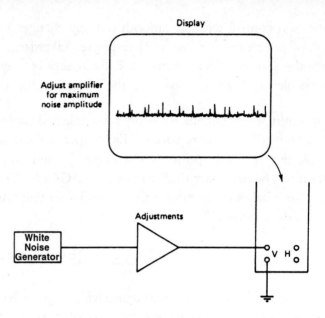

Figure 5-13. A white noise generator can be used for quick alignment of bandpass amplifiers and detectors.

DIGITAL

In *Figure 5-14a* write the output for each of the inputs (a-d). Then make a truth table. The truth table is shown in *Figure 5-14b*. It is made from the inputs and outputs in *a*. The procedure for writing this truth table can be used for all complex logic circuits.

SERIAL AND PARALLEL TRANSMISSION OF DATA

A multiplexer is a digital circuit that allows one of a number of different inputs to be selected, one at a time. *Figure 5-15a* shows the concept of a 5 to 1 multiplexer.

One application of a multiplexer is to convert parallel inputs to a single line. The various logic inputs are selected one at a time and the output is a series of logic levels spaced by the time it takes the multiplexer to go from one data input to the next.

The series data output of the multiplexer allows one line (for example, a telephone line) to transmit all of the data presented to the input of the multiplexer. This is called serial transmission of data.

For parallel transmission, all of the data is sent at the same time, see *Figure 5-15b*. Parallel data transmission is much faster, but it requires a separate line for every input of information.

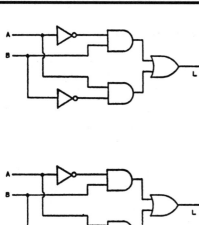
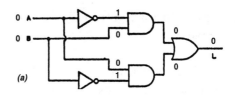

(a)

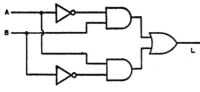
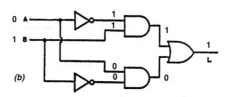

(b)

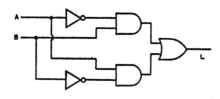
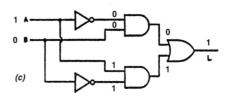

(c)

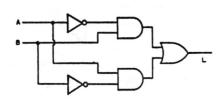
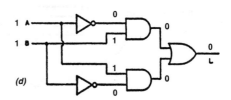

(d)

Figure 5-14a. Circuit inputs and outputs for writing the truth table.

Figure 5-14b. Circuit inputs and outputs with truth table.

Truth Table

A	B	L
0	0	0
0	1	1
1	0	1
1	1	0

Signal Injection and Signal Tracing   167

The output information from a microprocessor or a computer can be transmitted in either serial or parallel.

Figure 5-15c shows the principle of a demultiplexer. It shows serial input data converted to parallel output data. For example, serial data may be delivered to a computer on a telephone. The demultiplexer converts it to parallel data that is more useful to the circuitry inside the computer.

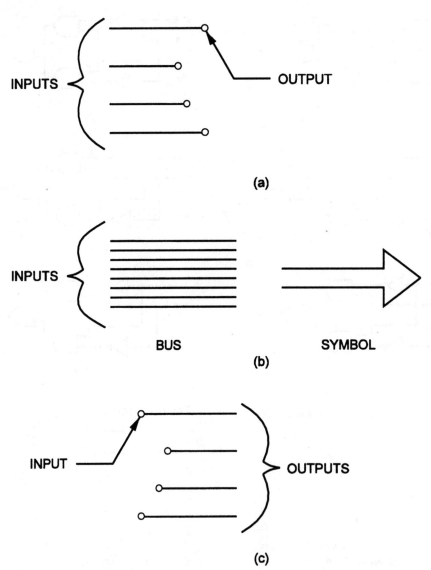

Figure 5-15. a) Operation of a multiplexer. b) The symbol for a bus. c) Operation of a demultiplexer.

SUMMARY

Two very valuable methods of troubleshooting any electronic system are signal tracing and signal injection. The methods are similar. You are looking for a section in the system that will not pass the signal. As a rule, the equipment needed for signal tracing is more complicated and expensive than that for signal injection. You should be able to use either method in troubleshooting.

In this chapter and elsewhere in the book, there have been discussions on how to make certain tests without using formal test equipment. *Do not be misled into thinking that test equipment is unnecessary.* While the quick tests are useful, the amount of information you get is very limited.

Using the oscilloscope in the frequency domain rather than the time domain is a very good way to check the frequency response of an amplifier. The manufacturer's literature will give specific details on how and where to inject signals. The slightly greater amount of time needed to set up a frequency domain test is well worth the effort.

Remember these precautions for frequency domain tests:

> Deactivate closed-loop circuits (such as AFC) before attempting a frequency domain test.

> Make sure the amplitude of the test signal is constant throughout the range of sweep frequencies.

CHAPTER 5 QUIZ

1. Troubleshooting a system means to

    A. isolate the defective circuit.

    B. find the faulty transistor.

2. Amplifier B in *Figure 5-16* is suspected of being defective. There is no output from amplifier C. What single component could be used to determine if the fault is amplifier B?

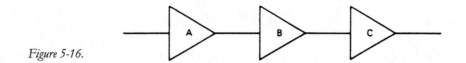

Figure 5-16.

3. Which of the following scope terminals is used for intensity modulation of the trace?

    A. Sync input

    B. Z1axis

4. Using an oscilloscope you are trying to trace a modulated RF signal. Your scope cannot reproduce the RF signal, but you can trace the signal by using a

    A. low-capacitance probe.

    B. demodulator probe.

5. Why is it a good idea to troubleshoot by starting in the middle of a system?

    A. To save time

    B. Because that is usually where the trouble is

6. The oscillator frequency of an AM radio is measured with a frequency counter. The frequency is 1095 kHz. The radio is tuned to

    A. 1095 kHz.

    B. 640 kHz.

7. Why is the axis of an oscilloscope used as a frequency marker?

   A. It doesn't distort the pattern.

   B. It is more accurate than adding the marker to the input signal to the scope.

8. The advantage of using a signal injector over using a screw driver is

   A. it provides a quicker method.

   B. it injects a wide range of signals for testing a variety of signals.

9. An AM receiver is suspected of having a dead oscillator. When the radio is tuned to 1240 kHz, what signal generator frequency should be substituted for the local oscillator?

10. Without resorting to the use of a sweep generator, a noise generator can be used to tune check amplifier and _____ circuits.

11. Noise signals that cover the audio and RF frequency ranges are called _____ noise.

12. Is the following statement true? When checking the frequency response of an amplifier, it is important to maintain the output of the amplifier constant throughout the test.

13. Name two types of signal generators that can be used for displaying the frequency response of an amplifier on a scope.

14. What is a disadvantage of using a sine wave voltage for sweeping the oscilloscope trace during a check for frequency response?

15. When displaying the frequency response of an amplifier, the oscilloscope is set up for a _____ domain display.

16. What simple tool can be used for signal injection?

17. The intermediate frequency for FM broadcast radios is 10.7 megahertz. If a receiver is tuned to a station at 101.4 megahertz, the local oscillator frequency should be _____ megahertz.

18. Which of the following waveforms would be better for signal injection?

    A. Sine wave

    B. Pulse

19. After preliminary checks, for quick troubleshooting, start

    A. at the signal input end of the system.

    B. at the signal output end of the system.

    C. in the middle of the system.

20. Incorrect alignment is a

   A. frequent cause of a system that doesn't work.

   B. rare cause of a system that doesn't work.

21. Make a truth table for the circuit in *Figure 5-17*.

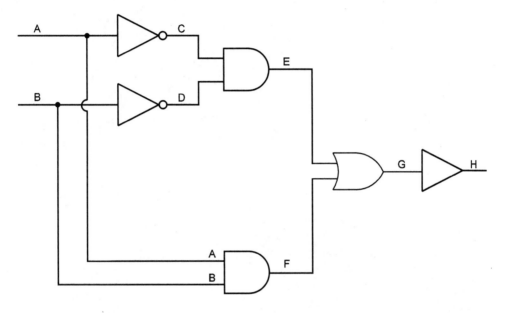

Figure 5-17. Is this circuit equivalent to an exclusive OR?

Signal Injection and Signal Tracing   173

CHAPTER SIX

Symptom Analysis, Diagnostics and Statistical Methods

Overview

It isn't likely that a single approach to troubleshooting and servicing equipment can be constructed. Technicians develop their own techniques; and even though they may seem cumbersome to another technician, they probably work very well.

Regardless of how well your favorite method of troubleshooting works for you, sooner or later you hit that brick wall. It is the tough dog problem that seems to defy all reason. When this happens, you need to convert to an alternate method such as:

> Take another look at the symptoms. Is it possible that some other section could cause that symptom?
>
> Is it time to consider the shotgun method? Replace every part, one at a time, in the circuit that you know is defective? There are cases where manufacturers actually recommend the shotgun approach.
>
> What do the records say? Your own records may be helpful. Also, there are computer-based records maintained by manufacturers and private companies. So, go for help.

Is there a flowchart or diagnostic available for the system you are working with? Consult the literature to see what is available.

Sometimes it helps to set the big problem aside and come back to it a little later. This chapter looks at some alternatives.

OBJECTIVES

What happens if the symptoms don't help?

What is a diagnostic?

What are the meanings of the shapes in a flowchart?

What is the order of statistical analysis?

Can a new battery be at fault?

SYMPTOMS

If there is any such thing as a common denominator in troubleshooting it would be to observe the symptoms. Even that can be disputed. A shop manager may say "start by doing the paperwork." But, for the purpose of this discussion, we will presume that the visual inspection and paperwork has been done.

You cannot rely on a description of symptoms from the customer, from a shop foreman or from any source other than your own observation. After all, as a technician, you are a trained observer and you can recognize types of symptoms better than the untrained.

There have been attempts to set up complete troubleshooting procedures based on symptoms only. In some cases that has been successful. For example, in a radio broadcast transmitter, the symptoms of impending failure are often indicated by taking readings from the various meters.

Symptom analysis may not work in cases where there is a large turnover of designs, such as television receivers. For example, technicians are trained to observe that no sound and no picture usually means trouble in the low-voltage power supply. That may seem like an obvious conclusion. After all, the sound section and the video section are both operated from the low-voltage supply.

However, if the low-voltage power supply in a receiver gets its energy from the flyback transformer, the trouble could be in the horizontal oscillator stage. In that case the memorized symptom would not be very useful. As a matter of fact, for every new design there are some possible new symptoms that would need to be learned for effective troubleshooting. So symptom analysis is very useful, but not as the only method of troubleshooting electronic systems.

As demonstrated by the example of the low-voltage power supply, it should be remembered that your knowledge of the system is very useful. Where there is no sound and no picture and no brightness, the question is: What part of the system supplies both the sound and video? In that case, knowledge of the system leads to the most obvious conclusion that it is a low-voltage supply or at least a point where DC voltages apply to both circuits.

OPEN THE CABINET

So far, none of the techniques have required the technician to touch the equipment. It has all been a mental exercise. Sooner or later it is necessary to open the cabinet and make some measurements and decisions based on those measurements. The first step in a hands-on troubleshooting procedure should always be to *measure the power supply voltage*. Nothing happens in electronic systems unless the power supply voltage is present.

If the voltage *is* present, but it is not the correct value, it can produce apparent troubles in any section of the system. As an example, suppose the audio section of a radio has a transistor-sensitive amplifier. This is an amplifier that produces a great amount of distortion if the supply voltage is not correct. The symptom would be *audio distortion*. The real problem could be in the power supply.

After measuring the power supply voltage and determining that it is correct, the next step depends greatly on the type of system involved. For example, in repairing a radio it is necessary to locate the troubled section, isolate the defective component and replace it within a few minutes. Otherwise, it would be cheaper to buy the customer a new radio than to suffer the financial loss that would occur by a technician spending too much time on a system that costs less than $100.

In a computer the next step would be to run a diagnostic that is on a floppy disk. In a tape recorder you can use visual observations to see if the tape is moving through the system and if it appears to be moving at the right speed. These are as valuable as the next step.

It is a good idea to look for hot spots, burn marks and places where a voltage arc-over may have occurred. All of those things are a part of your general routine when you start a troubleshooting procedure.

THE DIAGNOSTIC

A diagnostic is a step-by-step procedure for troubleshooting an electronics system. Diagnostics take on several forms. One is the basic *flowchart*. Symbols that are often used on flowcharts are shown in *Figure 6-1*. A typical flowchart is shown in *Figure 6-2*.

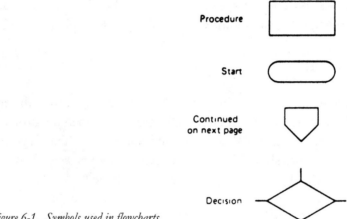

Figure 6-1. *Symbols used in flowcharts.*

The manufacturer leads the technician through a step-by-step procedure for isolating troubles in a television set. The diagnostic shown in *Figure 6-2* is typical. Note that the diamond-shaped boxes in the diagnostic represent a decision to be made, and the square blocks represent individual steps in the troubleshooting procedure.

This type of troubleshooting is very effective, but it has one drawback. If you have to start at the beginning every time you troubleshoot a piece of equipment, you are going to lose some time on problems that don't occur until late in the troubleshooting diagnostic procedure. In other words, you can often save time in some procedures by jumping ahead *if the symptoms indicate trouble in a specific section.*

Technicians in Kansas and Wisconsin reported in meetings that they run through the complete diagnostic procedure only once. From that time on they are sufficiently familiar with the equipment to troubleshoot it without resorting to flowcharts. However, if you are trying to fix one of those tough dogs, it would be a good idea to start at the beginning. If you are not sure where the trouble *is*, you can't be sure where it *isn't*.

Experienced technicians in those meetings claim that one of the most useful tools they have is the notebook they carry. They make notes on troubles that they have found in specific kinds of equipment. Their experience has been that troubles repeat. If you haven't seen a specific VCR for six months, it is easy to forget the cause of a particular symptom. So, after they locate it they jot it down in a notebook. The technicians claim that their notebooks are one of their most useful troubleshooting tools. That is always a good idea, but it requires a certain amount of self-discipline.

There is another good reason for keeping a personal notebook. Manufacturers may deny this, but there are weaknesses in many models of electronic systems. For example, an SCR may burn out frequently in one design, a power transistor may burn out often in another design, and so on.

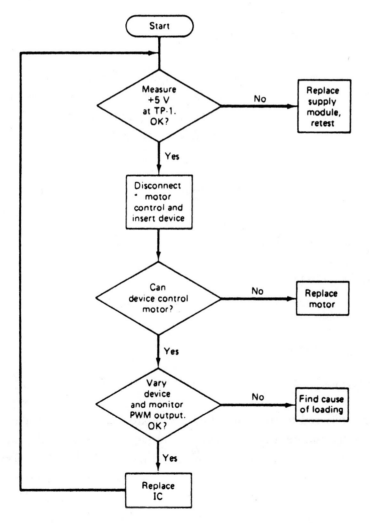

Figure 6-2. A typical flowchart.

Symptom Analysis, Diagnostics and Statistical Methods

When several technicians are working on equipment by the same company and on the same equipment, it is a good idea to compare notes. Also, keep an up-to-date file on company change notices. One of the advantages of belonging to technician organizations has always been the opportunity to exchange ideas.

Some companies provide a computer analysis for some types of equipment. For example, if you are troubleshooting television receivers you can subscribe to a computer system that tells you the most likely components for a specific type of failure. Subscription to that type of system can save many hours of frustration and will certainly repay in terms of efficiency.

Repairing computers requires familiarity with diagnostic programs that run the computer through its paces. These programs will tell you what specific section has failed in the diagnostic procedure. Diagnostics are well worth their cost and they are readily available.

STATISTICAL ANALYSIS

If you get to the point where you feel it is necessary to shotgun a circuit, it's smart to consider statistical analysis. With that approach, you replace components in the order that they are most likely to fail. Statistical analysis is not just for shotgunning a circuit. Many times the trouble will be traced to a specific circuit, and there are two or more components that can produce the symptom.

Consider the power amplifier circuit in *Figure 6-3*. It is conducting at near saturation. Here are some possibilities:

> There are shorted turns in the inductive output.
>
> Resistor R_1 is greatly reduced in value.
>
> Resistor R_2 is greatly increased in value.
>
> The transistor has an emitter-collector short circuit.
>
> Capacitor C_1 is shorted or very leaky.

If all of these components are soldered to the printed circuit board, you have to make a decision which to change first. This assumes that you don't have an in-situ tester like the one described in Chapter 11. Even if you have an in-situ tester, you have to make a decision about which remaining component is most likely to be the cause of the trouble. *Table 6-1* lists the components in the order of their most likely failure.

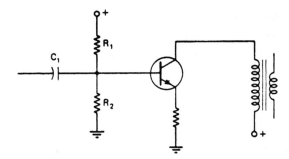

Figure 6-3. *A power amplifier used in the discussion of troubleshooting.*

For the example of the power amplifier, the most likely component to fail is the power transistor. A common fault with bipolar power transistors is an emitter-collector short circuit.

Unnecessary soldering and unsoldering components on a printed circuit board is a good way to introduce new troubles. Certainly, if the capacitor is defective the voltage at the base of the transistor will not be the correct value. So, before unsoldering any component in a power amplifier circuit, like the one in *Figure 6-3*, measure the emitter-to-base voltage and also the base voltage of the transistor.

Statistical Probability of Failure

Batteries

High-power devices
    Does it get hot in normal use? Diodes, transistors,
    ICs, tubes, ultrasmall electrolytics

Other transistors, ICs, diodes

High-power passive devices

Low-power passive devices
    Capacitors
    Resistors
    Inductors

Based on information from the Electronic Industries Association

Table 6-1.

If the manufacturer's schematic doesn't tell you what that voltage should be, you can make a quick estimate using the proportional method for finding voltages. It is an approximation but it should be sufficient to tell you if a leaky capacitor is a likelihood. The emitter-to-base short test for the transistor will tell you if the transistor is shorted internally.

To summarize, it is important to use measurements to determine the most likely component to fail. Having done that, and assuming the measurements have failed to identify the specific trouble, you can use *Table 6-1* to set the order of part replacement. You should modify that table so that it corresponds with your experience from working on specific systems.

A valuable troubleshooting tool is the in-situ tester that makes it possible to check the electrolytic capacitor and/or the transistor while it is still in the circuit. These in-situ testers are readily available.

As pointed out before, unsoldering components on a printed circuit board is not a good idea if you want to avoid working on the same system again in the very near future. In portable equipment it is usually a battery that serves as the source of voltage for the system. Many times the person squawking the equipment will say, "It can't be the battery. I just put in a new one." Unfortunately, new batteries can be defective.

Another problem you have to watch for is a customer or untrained personnel putting the battery in backwards. That works OK for flashlights (sometimes), but it can be highly destructive to transistor equipment.

SUMMARY

If you are troubleshooting familiar equipment, you will seldom have to refer to a flowchart or other kind of diagnostic. Even the tough dogs can be analyzed using standard troubleshooting procedures.

When everything else fails to get to the source of the trouble, it is best to turn to your personal notebook, the manufacturer's flowcharts and other aids (such as change notices and the shotgun method).

As a rule, the shotgun method is the absolute last resort. For large complicated systems, such as a personal computer, the diagnostic on a disk is the place to start. Those disks run the computer through a series of programs designed to check every section.

Technicians claim that one of their best aids is their own personal notebook. They use the notebook to write down peculiarities of different models of equipment they have serviced.

Special helpful hints can also be recorded in a personal notebook. They are printed in company bulletins, technical magazines and publications by technician organizations. Some companies provide computer access to data banks that store symptoms and cures for equipment problems.

A disadvantage of using a diagnostic or flowchart is that you always have to start at the beginning. The problem may not be anywhere near the starting point.

If you have to make a decision between components that may be causing the problem, remember the statistical chart. Replace the component most likely to fail.

CHAPTER 6 QUIZ

1. Is the following statement true? The best information a technician can use for troubleshooting is an eyewitness account of what happened when the system failed.

2. Is the following statement correct? Symptom analysis will always work as a method of troubleshooting.

3. If there is no sound and no picture and no screen brightness for a defective television receiver, a likely place to start would be the

   A. low-voltage power supply.

   B. sync separator.

4. After a visual inspection, what is the first step in hands-on troubleshooting?

5. A certain radio has a power supply voltage that is too low. Can this cause a symptom of sound distortion?

6. What visual observations are useful when you first start troubleshooting a defective system?

7. Is the following statement true? Sometimes there are weaknesses in the design of some equipment that cause repeated failures in certain models of equipment.

8. Which of the following components is more likely to fail in a system?

    A. Resistor

    B. Power transistor

9. Which of the following is more likely to fail?

    A. Rectifier diode

    B. Film-type capacitor

10. Is the following statement correct? The customer reports that new batteries have just been installed, so you should not waste time checking out the battery supply.

CHAPTER SEVEN

Servicing Closed-Loop Circuits

Overview

If the troubleshooting jobs are listed according to difficulty, then catastrophic failures would probably be the least difficult to troubleshoot. Closed-loop circuits and intermittents are among the most difficult.

Closed-loop circuits can be put into two main categories: those that control voltage and those that control frequency (or phase). They are especially difficult to troubleshoot when they are in discrete (individual component) form.

Examples of closed-loop circuits that control voltage are AGC/AVC feedback loops and regulated power supplies. Examples of closed-loop circuits that control frequency are the automatic frequency control (AFC) and the phase-locked loop.

Closed-loop circuits are difficult to service because their output depends upon their input, and their input depends upon their output. Any trouble anywhere in the circuit can change all of the voltages, currents and frequencies in the circuit. Fortunately, there is a standard procedure that works very well for both kinds of closed-loop systems.

Objectives

What is the standard procedure for troubleshooting all discrete feedback (closed-loop) circuits?

How is frequency synthesis accomplished?

What kind of voltage is usually used in both voltage and frequency closed-loop systems?

How does a closed-loop circuit affect the alignment of a receiver system?

Why isn't it correct to align a radio for maximum output?

What is the disadvantage of an analog voltage regulator?

What is a 555 timer and how does it work?

Closed-Loop Circuits for Voltage Control

The closed-loop circuits that control voltage will be discussed first. *Figure 7-1* shows the block diagram of a basic radio AVC loop. It samples the output signal voltage at the detector, filters it and delivers a DC gain-control voltage to early stages such as the RF and the first IF amplifiers. In other receiver systems, it is called automatic gain control (AGC). We will refer to it as the AVC/AGC circuit.

The purpose of the circuit is to keep the output sound signal at the speaker relatively constant, even though the strength of the incoming signal may vary. The variations in input signal may be caused by changes in atmospheric conditions and changes in the height of the earth's ion layer that affects long-distance reception.

The operation of the circuit is easy to understand. Suppose the input signal from the antenna increases in amplitude. This would normally cause an increase in sound volume from the speaker. However, part of the detector signal is rectified and filtered to produce the AVC/AGC DC control voltage. The resulting DC voltage is fed back to the RF and IF circuits to reduce the gain of those stages. So the increase in sound volume at the speaker has been prevented.

Suppose, on the other hand, that the input signal from the antenna decreases. That would normally decrease the sound volume at the speaker. However, the AVC/AGC DC control voltage now increases the gain of the RF and IF amplifiers in order to keep the sound volume at the speaker constant.

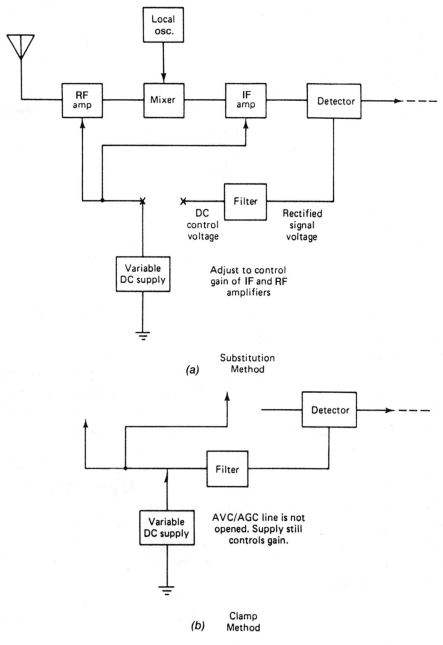

Figure 7-1. *Two methods of defeating the AVC/AGC circuit.*

To summarize, the AVC/AGC circuit produces a DC control voltage that holds the output sound volume constant by adjusting the gain of the RF and IF amplifiers.

As with all closed-loop systems, the difficulty in troubleshooting this one is the fact that the output from the detector depends on the amplification of the input signal, and the amount of a signal amplification depends on the amount of detector output signal. If there is trouble anywhere in the loop it is difficult to find because everything depends on everything else within the loop. There is only one sure way to troubleshoot this circuit effectively: defeat the loop. This can be done in several ways.

One way is to disconnect some part of the feedback loop and insert a DC power supply voltage that substitutes for the AVC/AGC DC voltage, see *Figure 7-1a*. Another way is to clamp the DC control voltage by placing a DC supply across the automatic volume control circuit. This method is illustrated in *Figure 7-1b*.

In either case, the gain of the RF and/or IF amplifier is controlled by the DC power supply. Once that is done, the standard troubleshooting procedures can be used. Clamping the AVC line, rather than opening the circuit and substituting a DC voltage, has the advantage of being quicker. It is not necessary to unsolder and resolder the AVC/AGC line, and that makes it a more reliable method.

Opening the loop has the advantage of making it possible to perform an important test of the circuit operation. By adjusting the power supply voltage in the opened-loop method, you are able to control the gain of the amplifiers. This, in turn, should change the amount of DC voltage available for feedback. If it doesn't, then the next step is to go to the output of the detector. Using an oscilloscope, look at the variation in detector input when you change the gain of the RF or IF amplifiers. If there is no change in the gain of the system when you vary the substitute DC voltage, then it follows that at least one of the amplifiers is defective.

The test setup in *Figure 7-1* is important for another reason. In some rare cases it is necessary to adjust or align the various RF and IF stages in a receiver. A common practice is to connect an AC voltmeter across the output in place of the speaker, see *Figure 7-2*. Then the adjustments are made for maximum output voltage as indicated by the meter. That procedure is actually given with some radio kits to enable the kit builder to adjust the receiver for maximum number of stations.

There are several problems with that procedure. The automatic volume control (or automatic gain control) affects the output.

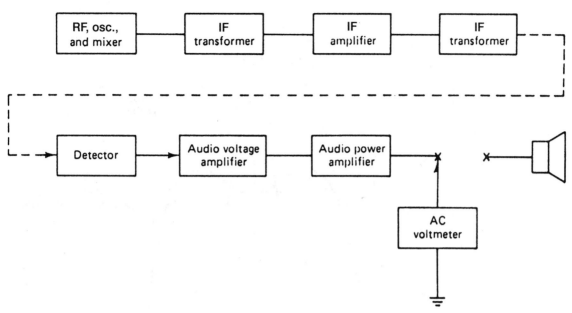

Figure 7-2. A measurement method sometimes used during alignment procedures.

Suppose, for example, you are adjusting a transformer between the IF stages. The adjustment, called alignment, is used to allow the receiver to pass only the IF frequencies and reject all others. As the alignment of the first IF transformer takes place, the amount of signal delivered to the IF amplifier is increased or decreased.

Suppose the alignment procedure produces an increase. This also produces an increase in the AVC/AGC feedback loop and reduces the gain of the receiver. Hence, increasing the gain of the amplifier by the adjustment, decreases the gain of the receiver by the AVC/AGC action. The overall effect of the AVC/AGC then, is to cause a broad adjustment of the transformer and an improper indication at the output using the method in *Figure 7-2*.

The second thing wrong with measuring the output audio during alignment as shown in *Figure 7-2* is that the maximum gain usually does not correspond with the maximum fidelity. To understand why, look at the illustration in *Figure 7-3*. Note the maximum signal amplitude passed through the transformer results in sharp skirt selectivity. So the range of frequencies, which in an AM radio should be about ±10 kHz, has been seriously decreased.

The effect then, of increasing the output as measured on the meter is to decrease the fidelity of the receiver because all of the useful information in an AM signal is in the sidebands.

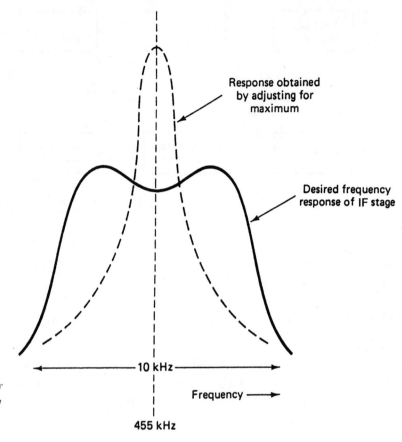

Figure 7-3. Comparison of maximum response with the desired frequency response of an IF stage.

Keep in mind that the same facts are applicable to FM receivers, communications receivers, television receivers, and in fact, all superheterodyne receivers. The message is clear: defeat the AVC/AGC system before any alignment is attempted.

Frequency Domain

The most effective way to align the IF stage or any other passband circuitry is to put the oscilloscope and signal generator into a frequency domain display and watch the passband while the adjustments are being made. *Figure 7-4* shows the test setup.

The sine wave generator may be built into the function generator. It is used for two purposes: it sweeps the function generator output frequency up and down from 450 to 460 kHz; and it moves the beam of the oscilloscope back and forth between the left and right side of the screen. For example, when the generator frequency is at 455 kHz the beam is the center of the cathode ray tube sweep.

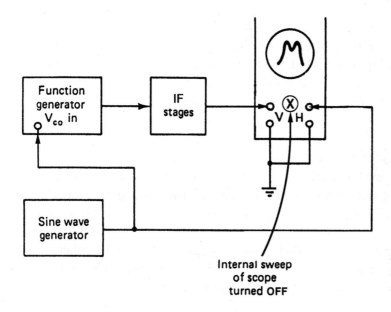

Figure 7-4. *Sweep method for looking at the IF response curve on an oscilloscope screen.*

THE AVC/AGC BIAS VOLTAGE

Alignment can be performed while looking at the AVC voltage, but this has no greater advantage over looking at the audio output at the speaker.

The bias voltage used to replace the AVC/AGC should be a pure DC. You can get close to this by using a variable 5V regulated supply, as shown in *Figure 7-5*. The problem is that there may be some ripple getting through the supply. Also, any noise on the line can pass through that type of setup. The overall effect of ripple or noise is to upset the alignment procedure.

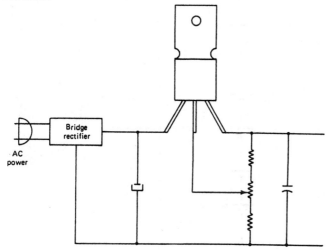

Figure 7-5. *A variable 5V supply.*

A better way is to use a battery pack, like the one shown in *Figure 7-6a*. At one time this had the disadvantage of being greatly dependent on the age of the batteries. So, over a long period of time, batteries had to be replaced. That was important because it could be a long time between uses. Today, the use of rechargeable Nicad batteries just about eliminates the problem of battery aging. Batteries can be periodically recharged as part of the calibration checkout period for the bench.

A very important advantage of the battery pack used for bias substitution is that it has a floating ground. In other words, there is no fixed ground for the output voltage. This means that it can be connected across any of the circuits without fear of grounding the circuit through a common connection in bias power supply.

The DC voltage used to substitute for the AVC/AGC voltage may need to be either positive, negative or both. The gain of a bipolar transistor amplifier can be controlled in two ways: forward AVC/AGC and reverse AVC/AGC. This is shown in *Figure 7-6b*.

In some complicated communications systems, the bias may be both forward (positive-going) in one stage and reverse (negative-going) in the next stage. The reason is that it increases the system stability. For example, if the RF amplifier bias is negative-going and the first IF amplifier bias AGC is positive-going, then the overall stability of the receiver is increased.

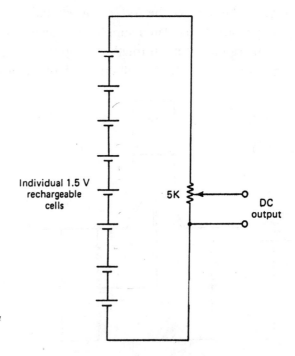

Figure 7-6a. Details of a battery pack circuit.

AVC/AGC systems are not only used in receivers. They are also used in hearing aids, public address systems, high fidelity audio systems and industrial electronic systems. Regardless of where you encounter this type of gain control, the message is clear: You cannot do an effective job of troubleshooting or aligning them unless you defeat the closed-loop circuit.

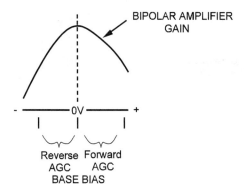

Figure 7-6b. The gain of a bipolar amplifier can be decreased with an increasing positive or increasing negative voltage on the base.

Analog (Linear) Voltage Regulators

In Chapter 6 a basic closed-loop regulator was discussed. *Figure 7-7* shows a block diagram of that system repeated here for your convenience. If you are going to troubleshoot this system, and the obvious preliminary methods don't work, open the loop and substitute a DC control voltage at some point in the feedback loop.

The logical place to open the loop is at the lead to the adjustable arm in the variable resistor. The point is marked with an X in the block diagram. This is the center lead on the variable resistor. Unsoldering the lead may not involve unsoldering a connection to the printed circuit board.

If the variable resistor is mounted to and soldered into the printed circuit board, the next most logical place to open the loop is at the base of the power amplifier. This point, marked with a Y, is easy to reach and is not usually connected directly to the board.

So, a new dimension is added to troubleshooting. In addition to the statistical method of choosing a possible faulty component, you need to pick components that can be investigated with a minimum amount of physical damage to the circuit.

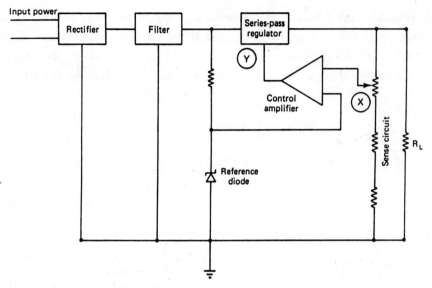

Figure 7-7. Block diagram of an analog voltage regulator.

Figure 7-8 shows a circuit using a three-legged integrated circuit (IC) voltage-regulated supply. This type of voltage regulation is very popular. The closed-loop circuitry is inside the package. So, if your troubleshooting procedure points to this device, it is usually a replacement job.

The indications of trouble in the three-legged IC are overheating, equal input and output voltages, zero output voltage and no supply regulation. Ripple in the output can be an indication of a faulty filter capacitor C or a bad regulator.

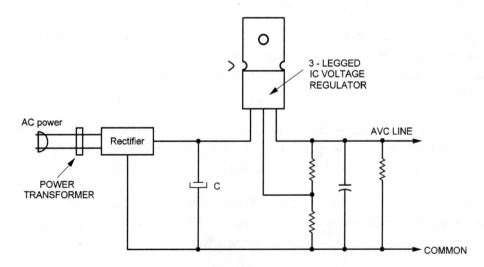

Figure 7-8. A fixed-voltage integrated circuit regulated supply.

194   Electronic Troubleshooting and Servicing Techniques

Switching Regulators

A disadvantage of a linear closed-loop regulator is that it is not efficient. A better way is to use a switching regulator; that is, if the most important consideration is efficiency.

Figure 7-9 shows the block diagram for a switching regulator. The relaxation oscillator supplies a pulse signal to the power control amplifier (controller). Regulation of the output is obtained by feeding back part of the output voltage and using it to control the width of the pulses. Wider pulses represent a high output power and corresponding higher output voltage. Because the frequency is high, the pulses are very easy to filter. So, less expensive and more effective filtering can be used with this closed-loop regulator.

To troubleshoot, start by opening the loop and substituting a DC control voltage. An oscilloscope is useful for troubleshooting this type of circuit. You can use it to look at the output waveform of the oscillator to be sure it is there, and that there is a change in pulse width corresponding to change in a DC substitution voltage at the sense input of the controller.

If you can't get the variation in pulse width, there is something wrong with the switching part of the circuit. If you can't get any output from the supply, you have probably already discovered that the oscillator isn't working. A frequency counter can be used to check the oscillator frequency if that frequency is known.

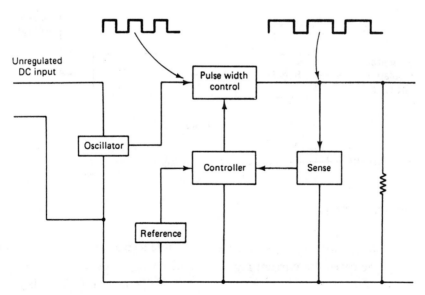

Figure 7-9. Block diagram of a switching regulator.

Servicing Closed-Loop Circuits 195

A specialized regulated power supply sometimes used in television receivers is illustrated in *Figure 7-10*. This scan-derived supply, gets its input power from a winding on the high-voltage (flyback) transformer. The flyback transformer is energized by pulses from the horizontal oscillator, which gets its energy from the scan-derived supply.

When this system is first energized, nothing can happen. The horizontal oscillator cannot oscillate because it is not receiving any power. So, there is no AC power delivered to the transformer and the regulator does not receive any power. This deadlock is broken by a start-up circuit. It is simply an oscillator that starts the circuit operation. After the circuit is started into operation, the start-up circuit is automatically disconnected from the system.

When you are troubleshooting a power supply like this, begin with the start-up circuit, which is easily identified on the schematic. An oscillator signal can be injected into the start-up circuit. This is the type of circuit that is best serviced by using a manufacturer's diagnostic chart.

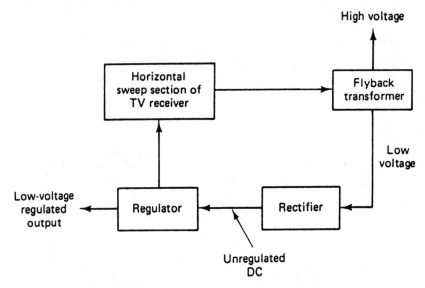

Figure 7-10. Block diagram of a scan-derived power supply.

CURRENT REGULATION

Instead of controlling power supply voltage, it is possible to use a feedback circuit to control the current. In some cases the current regulator and voltage regulator are both used in the same circuit. *Figure 7-11* shows a basic current regulator. Transistor Q_1 is called a series-pass regulator. It controls the current through R_1 and R_L. Control transistor Q_2 is forward biased by the voltage across R_1.

If the current through R_1 and R_L tries to increase, then the forward bias on Q_2 increases. Q_2 conducts harder and there is an increase in current through R_2. This decreases the forward bias on Q_1 and the current through R_1 and R_L is returned to normal.

A decrease in current through R_1 and R_L decreases the forward bias on Q_2. There is less current through R_2 and the forward bias on Q_1 is increased. This returns the current through R_1 and R_2, so the output current is returned to normal.

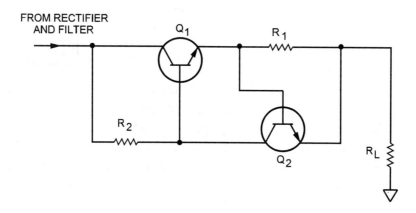

Figure 7-11. Basic current regulator.

Feedback Circuits that Control/Establish Frequency

Oscillators are circuits that employ regenerative (positive) feedback. In other words, the feedback signal is in phase with the input signal.

Figure 7-12 shows a block diagram of a typical sine wave oscillator. Note that an amplifier is an important part of the circuit. For this reason, you can often determine whether an oscillator is working simply by treating it as an amplifier. For example, start by measuring the emitter-to-base voltage and collector voltage if it is a bipolar circuit.

You cannot limit oscillator troubleshooting to this approach, however. The oscillator can be working, but it can be working at the wrong frequency. A much better approach is to use a frequency counter, which would tell you if the oscillator is working and also if it is working at the right frequency. A frequency counter is a very good investment for today's technician.

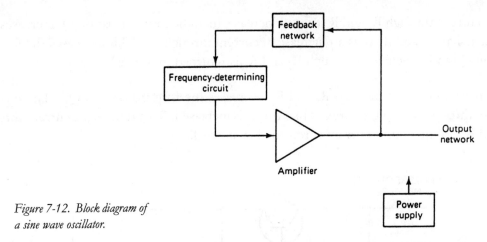

Figure 7-12. Block diagram of a sine wave oscillator.

Your knowledge of electronics is one of your best troubleshooting aids, and it is a good idea to have the block diagram of the oscillator clearly in mind. If there has been a catastrophic failure in the oscillator, the amplifier may still appear to be working properly. With vacuum tube and some FET oscillators, lack of oscillation can lead to complete loss of bias. Since it is a closed-loop circuit, it is not possible to tell whether the problem is in the amplifier, the feedback system or the other supporting systems. The indication of failure is that the frequency counter shows no output frequency.

An in-situ tester can be very important here. It will show if the bipolar transistor is OK and capable of amplifying and supporting oscillation.

Dip meters can be used to inject a signal into the oscillator-tuned circuit. If oscillation still doesn't occur, then it is a good possibility that there is a problem in the feedback circuit. In its simplest form, the feedback circuit uses a transformer or autotransformer. However, the feedback in other oscillators can also be a capacitive feedback circuit, see *Figure 7-13*.

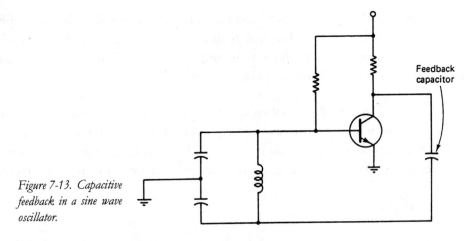

Figure 7-13. Capacitive feedback in a sine wave oscillator.

One good way to check oscillators is to open the loop and inject the required feedback signal into the amplifier. With the frequency meter you can then check to see if that signal is going all the way around the loop and back to the point where you opened it.

The tuned circuit is usually an LC circuit for sine wave oscillators, and an RC or RL circuit for relaxation oscillators. An exception to this is the phase shift oscillator, see *Figure 7-14*. The feedback network for this sine wave oscillator is an RC network.

Failure of the tuned circuit is not a common fault with oscillators. So, in a statistical approach, the amplifying device would be the first to suspect. This is one of the few feedback circuits where you cannot use a DC voltage to defeat the loop for troubleshooting.

In order to determine whether the oscillator is at fault for the failure of a system, you can disable the oscillator, usually by disabling the frequency-determining network. Then you can inject the required oscillator signal into the system at the output of the oscillator stage. If the system works properly, you have oscillator troubles and your next step is to troubleshoot the oscillator circuit.

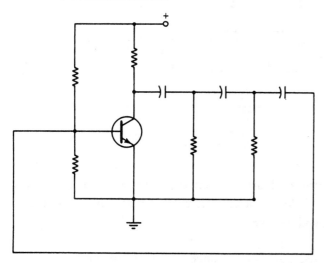

Figure 7-14. Another example of a sine wave oscillator using capacitive feedback.

AUTOMATIC FREQUENCY CONTROLS (AFC)

Figure 7-15 shows a block diagram of a typical AFC system. Its purpose is to lock the oscillator frequency to a specific frequency. At some point in the system the frequency is sampled. To give a specific example, in a television receiver the IF frequency, which should be about 41 MHz, is sampled for the AFC action. The sampled frequency is fed to an FM detector, such as a discriminator or ratio detector. The output of the FM detector is a DC voltage that is a direct function of the input frequency.

You usually see FM detectors used to demodulate an FM signal where the input frequency is varying according to an audio rate. In that type of circuit the output is an audio signal. However, in the AFC circuit the input is an IF frequency and the output is a DC voltage.

Assume that the input frequency is off the desired center frequency by a certain amount. That produces a DC output. The detector DC output is used to control a V_{co} frequency.

To troubleshoot this system, substitute the DC from the discriminator or ratio detector with the proper DC voltage. Vary the DC voltage and see if the oscillator frequency varies. (Use your frequency counter here.) If the frequency does not vary, your trouble is in the oscillator circuit. If it does vary, your trouble is in the feedback loop.

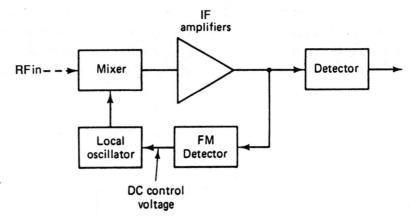

Figure 7-15. Block diagram of a typical AFC system.

THE PHASE-LOCKED LOOP

Putting the phase-locked loop in an integrated circuit opened the way to a wide variety of applications for this circuit in all fields of electronics. Rather than try to cover examples in all of the fields, two specific examples will be given.

Figure 7-16 shows a phase-locked loop used as a motor speed control. In one part of that illustration, a crystal-controlled frequency is used as the reference frequency. It is counted down before being delivered to the phase comparator.

The second input to the phase comparator comes from the feedback loop by way of the low-pass filter, amplifier, voltage-controlled oscillator, and frequency divider.

To start the discussion, it will be assumed that the input frequency from the reference, after countdown, is 10 Hz and the frequency divider is 1 — that is, the crystal-controlled frequency will be divided by 1.

If the two input frequencies to the phase comparator are the same, there will be a DC output from the frequency and phase comparator. It will be passed by the low-pass filter, then amplified (in some cases), and delivered to the voltage-controlled oscillator. The DC voltage into the VCO indicates that there is no frequency error. Therefore, the output of the VCO will be equal to the same frequency as the reference input.

If the VCO frequency is too high, there will be a DC voltage out from the phase comparator that is fed back through the loop. It will bring the oscillator frequency to the required value. Conversely, if the oscillator frequency is too low, the feedback loop will produce a DC voltage to the VCO that will raise its frequency. That will make it equal to the reference input to the phase comparator.

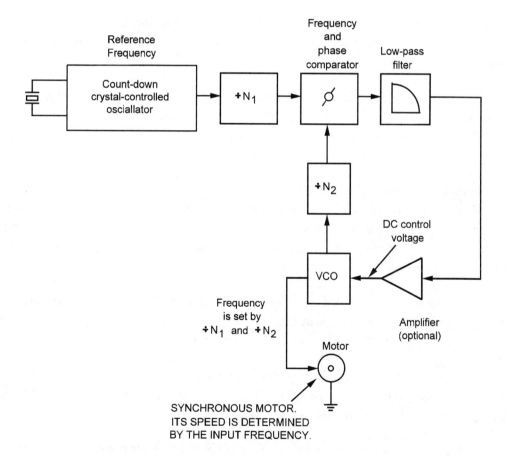

Figure 7-16. *A phase-locked motor control.*

Servicing Closed-Loop Circuits 201

So far, nothing really has been accomplished because the output frequency of the VCO is exactly equal to the reference frequency. Since the reference is crystal-controlled there is little point in using the phase-locked loop. But, suppose it is desired to change the output frequency to 100 Hz and still have it locked to a crystal-controlled reference signal.

In that case, the reference input will still be 10 Hz. The programmable divider will now be set to divide by 10. Now, in order for the two inputs to the phase comparator to be equal, it is necessary for the VCO to be operating at ten times the reference frequency, or 100 Hz. That way, when it is divided by 10, it will be equal to the reference frequency. So the output of the VCO will now be 100 Hz.

You can set the programmable divider to any division and change the oscillator VCO output to a desired value you want.

If the frequency divider in the reference circuit is also changeable, it is possible to produce any desired frequency (within the range of the system) that is crystal-controlled. This is an example of frequency synthesizing.

The motor speed control requires a motor that has a speed determined by the frequency of its input signal. One example is a stepping motor in which the number of pulses per second delivered to the stepping motor determine its RPM. Another example is a synchronous motor.

Frequency synthesizers of the type just discussed are used extensively in communications equipment and in consumer electronic equipment. As an example, the tuner in a broadcast radio can be controlled by a microprocessor that controls the programmable divider. In that way the microprocessor can set the desired master oscillator frequency for any specific broadcast frequency chosen.

Troubleshooting the phase-locked loop involves opening the loop if it is *not* an integrated circuit. A logical place to do that is at the input of the VCO. Replace the amplifier output DC voltage if it exists, or the DC output of the low-pass filter.

The amplifier is not always used. Its purpose is to make the feedback circuit more sensitive to very small changes in VCO signal. However, if the circuit is too sensitive, sometimes oscillations may occur. In that case, the amplifier is not used.

A variable DC voltage supplied to the VCO input should make it possible to determine if the loop is working. Change the DC voltage and see if the DC output (now disconnected) changes. If not, it is likely that the oscillator isn't changing frequency. Check it with a frequency counter as you change the DC voltage. Make sure that the frequency reference is correct using the frequency counter. If that doesn't show the problem, you can suspect the phase comparator.

In this chapter we have emphasized the use of a (digital) frequency counter. However, the skillful use of an oscilloscope can be used to measure frequency. The (digital) frequency counter is, by far, more convenient.

REVIEW OF SOME BASIC CIRCUITS USED IN CLOSED LOOPS

There are some basic circuits that are often used in frequency-control feedback systems. A few will be reviewed here.

VCO - Voltage-controlled oscillators have an output frequency that depends on the DC voltage delivered to their control electrode.

Discriminator/Ratio Detector - Both are circuits that have an output voltage that is dependent on their input signal control frequency. If the input RF signal frequency varies according to an audio signal, as in FM, the output of the circuits will be the audio modulating frequency. Remember, however, if the input RF frequency is unvarying, but is different from the center frequency of the discriminator/ratio detector, the output will be a DC voltage.

Phase Comparator - This circuit provides a DC output voltage that is directly related to the difference in phase between two input signals. It is sometimes combined with automatic frequency circuitry to make a circuit called AFPC (Automatic Frequency and Phase Comparator).

Programmable Counter/Divider - This is a digital circuit that performs either of two functions. It can count up or down. Also, it can be used to divide an input signal frequency by an amount determined by the control input.

Synthesizing - A term that means to make a new frequency by combining input frequencies.

OP AMP LOGIC COMPARATOR

Although it uses an analog operational amplifier, it is a valuable circuit in some digital systems. *Figure 7-17* shows how it operates.

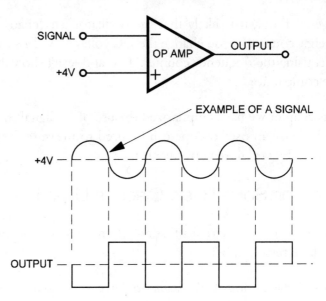

Figure 7-17. The operation of the op amp logic comparator.

The op amp comparator is used in an *open-loop* configuration. When the signal voltage on the non-inverting terminal is lower than the signal on the inverting terminal the output is high. When the signal voltage is higher than the positive terminal, the output is low.

The op amp is operated with a supply voltage of +5V. So the output swings between +5V (a logic 1) and 0V (logic 0).

Summary

There are two major categories of closed-loop systems: those that control voltage and those that control frequency. When discussing closed-loop control circuits, it is implied that negative feedback is being used. In other words, the feedback is used in some way to influence the output of the system.

Positive feedback is also used in systems and circuits. The best known use of positive feedback is in oscillators.

Returning to negative feedback control, both kinds (voltage and frequency) utilize a DC voltage for their operation.

Troubleshooting the closed-loop control circuits usually requires that the loop be opened at some point where there is a DC control voltage. A power pack or power supply is used to substitute for the DC control. Then the circuit can be treated as an open loop.

Instead of opening the loop, it is sometimes possible to clamp the DC control voltage with a fixed DC source.

Troubleshooting positive feedback oscillators is usually accomplished by taking measurements on the amplifier. A substitute oscillator frequency can be injected to determine if the frequency-sensitive networks are OK.

CHAPTER 7 QUIZ

1. What are two of the most difficult troubleshooting problems?

2. A high frequency is used in a power supply switching regulator because

    A. they are easier to troubleshoot.

    B. filtering is more easily accomplished.

3. Where does the AVC/AGC voltage come from?

    A. Radio power supply

    B. Detector stage

4. A battery pack is sometimes preferred for defeating the AVC/AGC voltage because

    A. it has no ripple.

    B. it is easier to connect.

5. You might find a start-up circuit in

    A. an analog power supply regulator.

    B. a switching power supply regulator.

6. Is the following statement correct? The purpose of an AVC/AGC circuit is to maintain a constant output of the radio under varying signal strength conditions.

7. Troubleshoot an oscillator circuit by

    A. opening the closed loop and substitute a DC voltage.

    B. testing the amplifier.

8. The IF transformers should be adjusted (if absolutely necessary) for

    A. maximum output signal.

    B. maximum fidelity.

9. Switching regulators control output voltage (and power) by controlling

    A. pulse width.

    B. pulse amplitude.

10. An AVC/AGC feedback voltage is

    A. always positive.

    B. always negative.

    C. sometimes positive and sometimes negative depending upon the system.

11. The statistical chance of a part going bad is a factor in component replacement. Another factor is _____ .

12. Name two types of feedback used in electronic systems.

13. Name two types of circuits that use degenerative feedback.

14. What kind of voltage is used for control in a degenerative feedback circuit?

15. What is the standard procedure for troubleshooting a closed-loop control system?

16. Name two kinds of closed-loop voltage regulator circuits for power supplies.

17. What is an advantage of a switching regulator over the analog type?

18. What circuit stabilizes output volume from a speaker despite changes in input signal strength?

19. In reference to the circuit described in Question 18, what is the DC control voltage used for?

20. Is the following statement correct? In an analog power supply regulator, the sense circuit is connected across the load resistance.

21. What type of component is used to get a reference voltage in a regulated power supply?

CHAPTER EIGHT

Hunting for the Causes of Noise and Intermittents

Overview

Although noise and intermittents are two separate types of problems, they are included in this one chapter because some of the methods used to track down the two problems are the same. Furthermore, intermittents can cause noise so they are also related in that way.

As mentioned in the Preface, it is very difficult to categorize some of the test procedures used by technicians. In some cases, the categories are strictly arbitrary. Part of this chapter is devoted to identifying types of noise and types of intermittents.

Objectives

How is noise related to the strength of an incoming signal in a receiver?

Where is the noise generated in a receiver?

What causes snow and hissing noise in receiving systems?

What do the names *Boltzmann's constant* and *Avogadro's number* mean to an understanding of noise?

How is noise related to the bandwidth of an amplifier?

What is 1/f noise?

ANALOG

RESISTOR NOISE

This discussion begins by considering a perfect amplifier, as shown in *Figure 8-1*. The perfect amplifier increases the amplitude of an input signal, but it does not add any noise to the signal output.

The output power, measured by the wattmeter, will show no power output when there is no input signal. Contrast that with an actual amplifier. With a real amplifier, the wattmeter would show some output power without an input signal. That output power would be due to noise generated within the amplifier.

The concept of the perfect amplifier is actually used to define noise measurements. So it is not just a hypothetical case that has no real meaning.

In *Figure 8-2* a resistor has been connected across the input of the amplifier. Now the output wattmeter shows a measurement, *even though the theoretical amplifier still does not produce any noise*. The output measurement is called *noise power*. It is the result of the amplifier increasing the amplitude of the noise generated by the resistor.

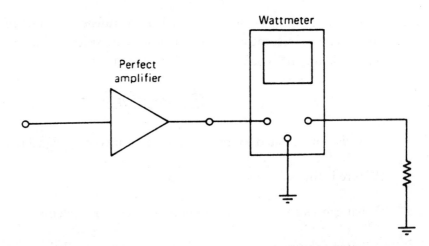

Figure 8-1. The idea of a perfect amplifier.

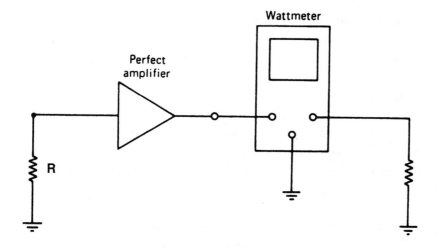

Figure 8-2. Connection of the resistor across the input terminals causes noise to be injected.

There is no current flowing in resistor (R) in the circuit. Obviously, then, the noise is not created by electrons "bumping into atoms and other electrons" as they move in a current flow. Where, then, does the noise come from?

To understand it, consider the model of a resistor shown in *Figure 8-3*. There is no voltage applied across this resistor by an outside source of power. At room temperature (and, in fact, at all temperatures above absolute zero) the atoms in the resistor are in continual motion called *Brownian motion*. At room temperature some electrons escape from the atoms for a short period of time. They flow through the material until they are captured by another atom that has lost an electron. The motion of electrons for a very short duration of time is called *intrinsic current*.

The motion of the electrons is random. However, at any specific instant of time there will be more electrons moving in one direction than in the other. At that instant of time a voltage is created across the resistor. Remember, the electrons are flowing because they have obtained enough energy to escape from their atoms in motion.

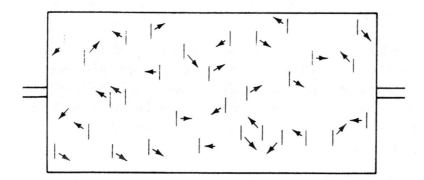

Figure 8-3. A model of a resistor at room temperature showing the random motion of electrons.

If you increase the temperature of the resistor there will be more electrons moving in the intrinsic current. Whenever the net amount of intrinsic current flowing in one direction is greater than that flowing in the other direction there is a voltage across the resistor. The higher the temperature the higher the amplitude of the voltage. That is because at a higher temperature there is more intrinsic current.

Over a period of time the voltages created by the random electron flow in the resistor results in a random fluctuation of voltage at the input of the amplifier in *Figure 8-2*. That random fluctuation is amplified by the (perfect) noiseless amplifier and creates the output noise power.

There is an immediate practical applications to the noise that has just been discussed. Certainly, most amplifiers have resistors at their input terminals and those resistors will produce noise. Resistors are something to consider when tracking amplifier noise. That noise can be very troublesome in some systems. If the noise is greater in amplitude than the incoming signal, then the signal-to-noise ratio is very unsatisfactory.

You must be very careful when you replace resistors in certain circuits. For example, what happens if you replace a film resistor at the input of an amplifier with a carbon composition resistor (which creates more random noise voltage)? The answer is that you will likely introduce enough additional noise to change the signal-to-noise characteristic of the system.

Good technicians are very careful to use exact replacements of resistors — especially for those parts of the system near a signal source.

This discussion could be applied to the input circuit of any industrial analog system or digital system. The input might be from a transducer (often called a sensor). The system is used to amplify the relatively weak transducer input to a higher and more useful amplitude.

As with the case of the television receiver, the noise created by resistance across the input terminals of any amplifier can render it useless. That is especially true if a careless technician replaces a resistor with the wrong type.

Here is another practical application of what has just been discussed: A well-designed communications receiver has just been installed in a new location and the technician connects the antenna terminals to an outside antenna. Immediately *the technician has injected noise into the system because of the antenna resistance.* Some types of antennas introduce more noise than others, and the technician should make sure to determine which antennas produce the highest possible signal-to-noise ratio in the system.

Remember this important point: Any resistor, conductor or semiconductor device connected across an input of a small-signal amplifier (that is, an amplifier that has high voltage gain) will introduce noise into that amplifier. *When looking for the source of noise, you should start by going to the source of the signal.*

Of course, if an amplifier has been working for some time and suddenly it begins to produce output noise the noise is not necessarily caused by a change in the input resistance. However, that is always a possibility that should be considered. Another likely cause is that the amplifier gain has been reduced. That, in turn, can change the signal-to-noise ratio to an undesirable value.

WHAT IS BOLTZMANN'S CONSTANT?

Whenever you see an equation for noise in an electronic system, you are sure to find the letter k — Boltzmann's constant. Good technicians don't like to work with things they don't understand, so a brief discussion of Boltzmann's constant will be given here.

Amedeo Avogadro (1776-1856) was an Italian physicist who startled the world of science with the following declaration: *Equal volumes of gases at equal temperatures and equal pressures have exactly the same number of gas molecules regardless of the kind of gas.* As you will see, it also applies to solids like carbon.

That wasn't much more than a guess when he first proposed it, but it turned out to be a very good guess. It has been verified a number of different ways by scientists. By defining the volume, pressure and temperature, the number of molecules was established to be 6.02×10^{23}. That is called Avogadro's number. Avogadro's guess is now called Avogadro's law. It was not immediately accepted, but eventually became a very important part of modern theory.

There is another very important law that can be stated as a basic equation:

$$pv = RT$$

If you multiply the pressure of a gas (p) by its volume (v), you get the temperature of the gas (T) (measured in degrees Kelvin) multiplied by a constant R. The value of R is the same for all gases, and it is called the gas constant. Degrees Kelvin is the number of degrees above absolute zero temperature.

$$\text{Degrees Kelvin} = 273° + \text{Degrees}° \text{ C}$$

You can determine the value of R by multiplying the pressure and volume of any gas and then dividing the product by the temperature of the gas. No matter which gas is used, you always get the same value.

Two very important constants have been defined: the gas constant (R) and Avogadro's number (N). Neither is equal to zero, so you can divide one by the other and get a third constant (k):

$$k = R/N$$

That value (k) is Boltzmann's constant. It is used in many scientific equations.

You will often see Boltzmann's constant as a multiplier for absolute temperature (T). That makes kT a unit of energy. It is used, among other things, in discussions of energy levels of atoms in materials.

The amount of noise generated by a resistor increases with temperature. That's because the atoms have more energy at higher temperatures. The energy levels are expressed as kT. So why does kT show up in equations for noise in transistors, resistors and so on? Because the amount of noise generated in those devices is directly related to the energy of the atoms (and electrons).

Using Boltzmann's constant, scientists are able to find out the expected electron energy and, hence, the amount of noise that will be created in various components. It is interesting to look at the equation for the amount of noise generated by a resistor.

The equation is not given so that you can, armed with a calculator, determine the amount of noise voltage in various parts of the equipment. As in many cases, an equation serves technicians only one purpose: to show you the relationship between some of the variables in the system. Here is the noise power equation:

$$\text{Noise Power} = kTB \text{ Watt}$$

In this equation, k is Boltzmann's constant, T is absolute temperature (in degrees Kelvin), and B is the bandwidth (in Hz).

The equation for noise voltage generated in a resistor is:

$$\text{NOISE VOLTAGE} = \sqrt{4kTBR}$$

Where kT and B have the same meaning as in the noise power equation, and R is the resistance of the resistor.

Various names have been given to the noise generated by a resistor at room temperature: *Johnson's noise, thermal noise, thermal agitation noise, or white noise.*

The term *white noise* is a misnomer. It comes from the idea that the noise signal produced by the random motion of electrons has a random frequency and amplitude. It is supposedly the same random relationship as the color white, which is supposed to contain all of the colors of the rainbow spectrum.

However, white is not white — at least as it is perceived by a human being. You could not make something look white on television by mixing equal parts of primary colors (red, green, and blue). But the name *white noise* persists. Think of it as a very wide spectrum of frequencies and amplitudes created by the random motion of electrons.

The equations just given clearly show the effect of temperature on the noise created in a system.

AN EXAMPLE OF NOISE

To get a direct idea of the effect of connecting a resistor across amplifier terminals, consider this example. A 5-megohm resistor connected across an amplifier that has a bandwidth of 5 megahertz will inject about 10 microvolts of noise into the amplifier at room temperature.

Ten microvolts can be an appreciable part of the total input of an RF amplifier. In some systems the total amount of useful input RF signal is about 50 μV delivered to the antenna terminals. So a noise signal of 10 μV would account for about 20% of the total input to the antenna terminals.

The equation also shows that *the amount of noise can be increased by increasing the bandwidth of the amplifier.* As you would expect, increasing the resistance of the resistor will also increase the amount of noise injected.

Johnson's noise is not the only noise you have to worry about in an amplifier. Refer to *Figure 8-4*. It shows the types of noise and their distribution in a typical electronic amplifier. Note that white noise has a constant amplitude throughout the bandwidth.

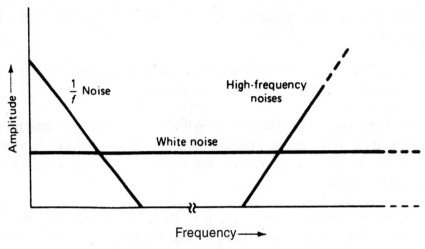

Figure 8-4. Three kinds of noise.

When current flows through a resistor, very small instant-to-instant changes in current cause a varying noise voltage. That is a source of 1/f noise.

If you heat the cathode of a vacuum tube, electrons are emitted from its surface. Unfortunately, they don't all leave like soldiers marching in unison. At any instant of time the number leaving the cathode will be different from the number at any other instant. That produces a special kind of noise referred to as *flicker noise, or 1/f noise*.

You might think you get away from flicker noise with transistors and field effect transistors. However, the cause of that type of noise in semiconductor devices is thought to be the flow of charge carriers across the surface of (rather than through) the device. It is still a form of noise.

Flicker noise is present in all amplifying devices. Unfortunately, there isn't much you can do about it, but it does decrease in amplitude as the frequency increases. Flicker noise is primarily a problem at the low-frequency end of the spectrum.

Another problem with amplifying devices is called *partition noise*. It can best be described by using the model of a bipolar transistor, see *Figure 8-5*. Note that of all the electrons leaving the emitter area, only a small part go to the base and the rest go to the collector region. The problem is that from instant to instant the number that go to the base vs. the number that go to the collector changes in a random way.

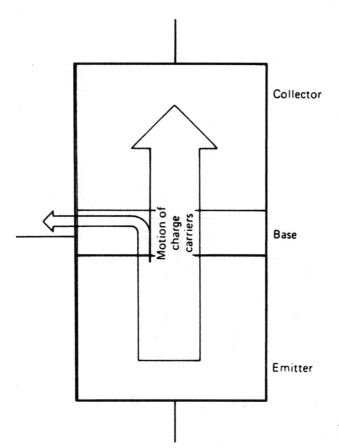

Figure 8-5. The cause of partition noise is illustrated in the bipolar transistor model.

This means that the base current is changing from moment to moment by a very small amount. The variations in current are a source of transistor noise. The small base current changes are amplified due to the transistor beta. So the amount of amplified output noise can be an important factor in the total amount of amplifier noise.

Partition noise is also a problem in vacuum tubes. In *Figure 8-6* you see electrons sticking to the control grid of a triode vacuum tube and returning to the cathode through the grid resistor. In a manner similar to the bipolar transistor, the number of electrons that actually flow through the grid circuit is subtracted from the output (plate) current. So there is a moment-to-moment change in current through the plate resistor.

Moment-to-moment variations in the partition current flowing through R cause a noise voltage to appear across that resistor. The noise voltage is amplified by the tube. Keep in mind the fact that the resistor is also a noise source.

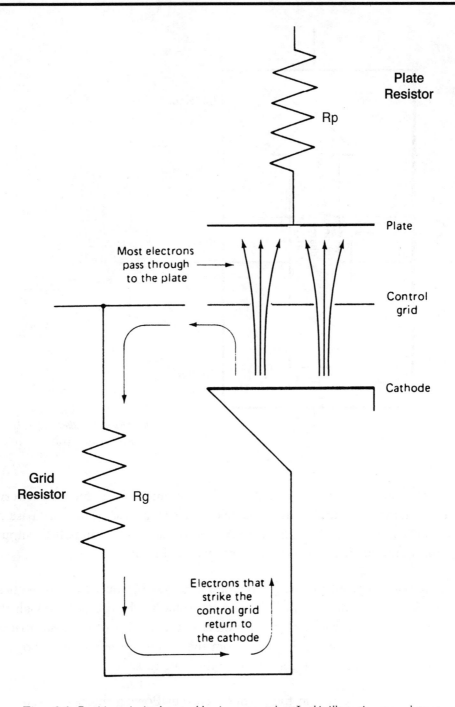

Figure 8-6. Partition noise is also a problem in vacuum tubes. In this illustration, some electrons return to the cathode through the grid resistor. The number varies from moment to moment.

In a JFET or MOSFET there is no motion of electrons into the gate region. That means that there is no partition noise in those devices. That is the reason why you see field effect transistors used extensively in modern communications and other high-frequency electronic systems, especially in RF circuits.

There are other kinds of noise generated in electronic systems, but the most important kinds have been discussed as far as the noise generated within the system is concerned.

EXTERNAL NOISE

Noise generated outside the system is very often some form of impulse noise. Examples include static created by lightning and noise generated by machinery. Impulse noise, as with any other pulse, contains a very large number of harmonic frequencies. It is a great disturbance in electronic systems.

Not much can be done about impulse noise for a given system. Some intelligent use of antenna theory may help. Coaxial cable prevents the impulses from being picked up by the transmission line. The shield prevents the line from acting like an antenna.

Impulse noise is usually horizontally-polarized, so long horizontal runs should not be used in a receiver antenna system.

AN EXAMPLE OF A NOISE PROBLEM

Now consider a system that was working OK at one time and all of a sudden requires servicing because of internal noise. This may be due to a transistor that has had a beta change that reduces the signal-to-noise ratio. It could also be due to a bad component, such as a resistor, which is becoming too hot.

The question is: How do you find the noise source? It is assumed that all preliminary tests have been done and the power supply voltages measured. (A low-power supply voltage can cause a reduced amplifier gain and a reduced signal-to-noise ratio.) Start troubleshooting noise problems by defeating all closed-loop circuits in the system. If, for example, the system has an automatic gain control, it is important to defeat that circuit.

Automatic frequency controls can also carry noise signals in a feedback loop because of the wide bandwidth of the noise. So it is necessary to defeat the automatic frequency control as well as the automatic gain control. The best way to do that is to replace the DC voltage in the feedback systems with a DC supply voltage — preferably from a battery pack that will not add hum or noise.

Receivers are not the only electronic systems that employ negative feedback. You will find negative feedback in industrial speed controls, hearing aids, public address systems and a wide variety of other systems. In all cases, those feedback circuits must be located and defeated when looking for the source of noise.

Having defeated all of the closed-loop systems, begin to look for the noise source by *starting at the input of the system*. For this type of problem, the idea of starting in the middle of the system should *not* be used. Remember that noise is created at the input end of the system, so there is no use in wasting time in the middle.

In an industrial system, start at the transducers; in a communications system, start at the antenna terminals; in an audio system, start at the microphone. Again, as shown by those examples, start at the input.

Moving away from the input, disable the circuits one at a time. If the noise stops, then the noise source is in front of the circuit you just disabled. Disabling a circuit usually means to ground the signal input or, in the case of bipolar transistors, to short the emitter to the base.

As you move through the system, listen to the output noise. This is best done by turning the system gain to the maximum value, and at the same time defeating the input signal from transducers, antennas or other signal sources. When you move past the offending circuit, the amount of noise in the output should be much lower.

Another method of finding noise is to think of the undesired noise as being an injected signal. Use the signal tracing method to find the source. Start in the middle of the system and work toward the input. When you pass the source of the noise (going toward the front end) you will no longer be able to detect it. This tells you where it is located.

INTERMITTENTS

Intermittents can be very frustrating troubleshooting problems in electronics. Think of the intermittent as being a switch that turns a signal on and off in the system; your job is to locate the switch.

The problem can be a break in the printed circuit board, a cold solder joint, a defective tube, transistor or FET, a broken wire or an intermittently shorted component. In fact, the variety of sources of intermittents is almost endless. An important thing to remember is that intermittents, like noise sources, will be very difficult to find if you do not defeat the closed-loop feedback circuits within the system. This is the first thing to do after performing preliminary checks and measuring the supply voltage.

Keep in mind that the power supply can be the source of the intermittent. So, it is a good idea to disconnect the power supply and substitute a battery pack. Look at the system output on an oscilloscope to see if it is constant.

Starting at the input, as in the case of noise, look and listen for the intermittent as you move through the system. While doing this, it is a good idea to slightly flex printed circuit boards and move components with an insulated stick to try to stimulate the intermittent.

Intermittents can be temperature related, so technicians often use a heat source (such as a hand-held hair dryer) to heat different sections of a circuit as a method of inducing or stimulating the intermittent action. This is especially useful in a type of intermittent that will only occur at higher temperatures. For example, when the system is in its cabinet, it will display intermittent behavior, but when it is out on the workbench, the cooler ambient temperature prevents the intermittents from occurring.

Technicians have to make their own decision about spraying supercoolants on components to stop noise or intermittent interference. They cool the component with a freon spray. That stops the intermittent if it is the type that occurs with a high temperature.

If you decide to use these supercoolants, remember this: *They can be very dangerous because they can produce instant frostbite! Furthermore, if inhaled into the lungs, they can produce permanent lung damage!* Many technicians avoid them. Others are prepared to exercise the safety measures necessary for their use.

DIGITAL

Clock Circuits

In computers and control systems it is a common practice to use a square wave signal as a reference for all other circuits. The square wave is called a *clock signal*, and it often comes from a multivibrator circuit that is labeled *clock*.

Remember that the NOT gate is actually an amplifier operated at either cutoff or saturation. In a multivibrator circuit made with two transistors, one is in saturation and the other is cutoff. On each half cycle, the operations of the transistors reverse.

It seems logical that an NOT gate can be used as a multivibrator. In fact, you will find such circuits used for a clock, see *Figure 8-7*. A tip-off that this is a multivibrator is in the use of a capacitor for feedback.

If a highly accurate clock frequency is needed, as in the case of clocks for telling time and stop watches, a crystal may be used in the feedback circuit.

When you are troubleshooting logic circuitry it is a good idea to check the clock signal. A logic probe can be used to show that a pulse is present, but an oscilloscope may be needed to study waveforms in some troubleshooting problems.

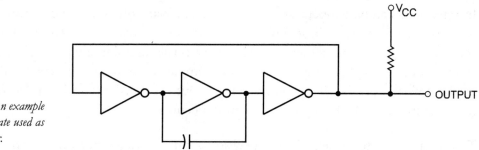

Figure 8-7. An example of an NOT gate used as a multivibrator.

Combined Circuits on an Integrated Circuit Chip

There are a great number of circuits made of gates and packaged in integrated circuit form. It would not be possible to list them all in a single chapter. When you get a book of IC logic pinouts it would be a good idea to look through it to see what is available.

An example of an IC package is shown in *Figure 8-8*. The quad NANDs in that illustration are contained in TTL package #7401. As another example, the SN74279 is a TTL package of four RS flip-flops.

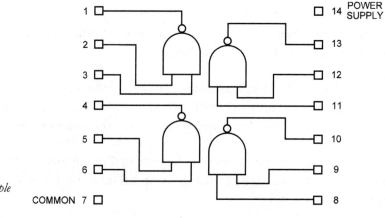

Figure 8-8. An example of an IC package.

222   *Electronic Troubleshooting and Servicing Techniques*

JK Flip-Flops

The RS flip-flop has a *not allowed* condition which could make it useless in some applications. The JK flip-flop has no such disallowed input. It is made by taking basic RS flip-flops and combining them with some other basic gates. The overall result is a more elaborate, but also more useful circuit.

Figure 8-9 shows a typical JK flip-flop. (Normally, two or more such flip-flops are located in a single integrated package.) If you replace the S and R with J and K in the flip-flops of *Figure 8-10*, the operation would be the same. However, there are additional terminals, as shown in *Figure 8-9*, that also affect the operation. There are JK flip-flops in all logic families. The discussion here is for a JK flip-flop in the TTL family.

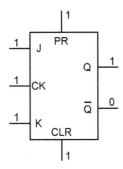

Figure 8-9. Normal voltages shown for the flip-flop in the high condition.

The CK terminal is used for a clock input signal. The JK flip-flop cannot change by switching J or K to logic 0 unless the CK terminal is switched to 0 at the same time, see *Table 8-1*. When K is switched to 0, but CK is held at 1, the flip-flop condition does not change. However, when the CK terminal is switched to 0 and K is switched to 0, then the condition of the flip-flop changes to low. This arrangement assures that the flip-flop is not accidentally switched with a noise signal, or with a glitch.

The 5th and 6th rows of the truth table in *Table 8-1* shows that the condition of the flip-flop changes when the CK terminal goes to 0. This is indicated by the symbol **. The SET and PRESET terminals are held at logic 1 for this operation.

The PR (preset) terminal is used to switch the flip-flop to high. This is shown in the first row of the truth table. This makes it possible to switch all flip-flops in a system to high at the same time.

The CLR (clear) terminal is used to switch the flip-flop to a low condition regardless of what condition it is in. The second row of the truth table shows this. The reason for having this operation is that there are times when you want all of the flip-flops in a system to switch to low at the same time.

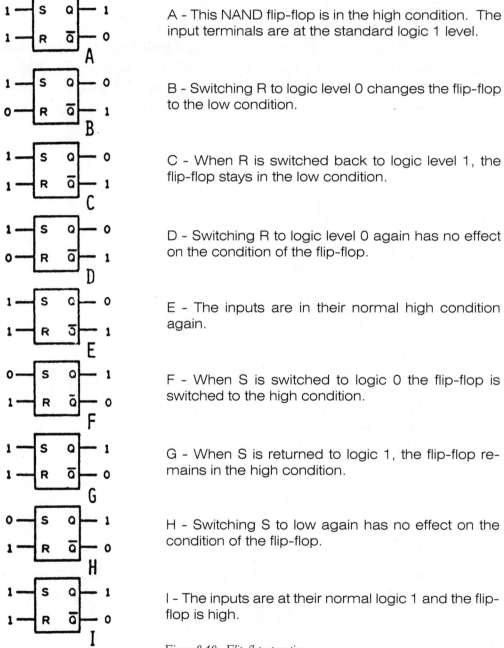

A - This NAND flip-flop is in the high condition. The input terminals are at the standard logic 1 level.

B - Switching R to logic level 0 changes the flip-flop to the low condition.

C - When R is switched back to logic level 1, the flip-flop stays in the low condition.

D - Switching R to logic level 0 again has no effect on the condition of the flip-flop.

E - The inputs are in their normal high condition again.

F - When S is switched to logic 0 the flip-flop is switched to the high condition.

G - When S is returned to logic 1, the flip-flop remains in the high condition.

H - Switching S to low again has no effect on the condition of the flip-flop.

I - The inputs are at their normal logic 1 and the flip-flop is high.

Figure 8-10. Flip-flop operation.

PR Preset	CLR Clear	Clock	J	K	Outputs Q	Not Q
L	H	X	X	X	H	L
H	L	X	X	X	L	H
L	L	X	X	X	H*	H*
H	H	**	L	L	Q_0	Not Q_0
H	H	**	H	L	H	L
H	H	**	L	H	L	H
H	H	**	H	H	Toggle	
H	H	H	X	X	Q_0	Not Q_0

H = High level (steady state)

L = Low level (steady state)

X = Irrelevant

** = Transition from high to low level

Q_0 = The level of Q before the indicated input conditions were established

Toggle = Each output changes to the complement of its previous level on each active transition (pulse) of the clock

* = This configuration is nonstable; that is, it will not persist when preset and clear inputs return to their inactive (high) state

Table 8-1.

The fourth row shows that when both inputs are switched to low the flip-flop does not change its condition. Remember that this is a *not allowed* condition for the RS flip-flop.

The toggle condition, which is the 7th row of the truth table, shows that if all terminals except the clock input are held at logic 1, then the flip-flop will change condition every time the CLK terminal is switched to 0. The CLK terminal is then called a toggle.

Schmitt Trigger

Figure 8-11 shows the logic symbol for a Schmitt trigger. Basically, the Schmitt trigger is used to produce a square wave output for almost any input waveform. A sine wave input will produce a square wave output. The TTL 74132 integrated circuit contains four Schmitt triggers.

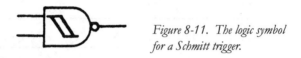

Figure 8-11. The logic symbol for a Schmitt trigger.

Types of Noises and Their Causes

It is important to understand that *most* of the noise created internally in a communications receiver is created in the front end. For example, in a television receiver, almost all of the troublesome noise generated in the receiver is created in the television tuner. The noise shows up as snow on the picture and it causes a hissing noise in the speaker.

In a system designed to communicate audio, the noise shows up as a hissing sound in the speaker. In other systems, such as digital and microprocessor, the internally generated noise is seen as grass on an oscilloscope display of the signal. Also, it can cause undesired triggering.

Do not confuse noise that is created internally by circuitry with the noise that is created outside the system. Using television as an example, atmospheric interference such as lightning creates noise that is generated outside the receiver

The amount of snow on television screens, due to internally-generated noise, is indirectly related to the strength of the incoming signal. A strong signal can override noise voltages. If the signal-to-noise ratio is high enough, the noise problem is eliminated.

If the signal is weak, the AVC/AGC system will raise the gain of the receiver. This also raises the system noise. System noise is the reason a good antenna system is needed in weak-signal areas. Hunting down noise sources and eliminating them may involve going outside the system to improve the strength of the signal.

SUMMARY

Much of the noise in a system is generated by resistors, diodes, transistors and other semiconductor devices. The amount of noise depends on some very basic theoretical considerations.

Avagadro's Law and Boltzmann's constant were originally related to the laws of gases. However, it is now possible to use those laws to calculate the amount of noise to be expected from a resistor or other semiconductor device.

A perfect amplifier does not add any noise to a system. A resistor connected across the input terminals of a perfect amplifier will result in a noise power at the amplifier output. So noise can be added by adding semiconductors at the input.

More important, replacing a semiconductor device with the wrong type can produce an increase in system noise.

Almost all of the noise in a system starts at the signal input end.

Before chasing noise or intermittents through a system, disconnect all closed-loop circuits.

CHAPTER 8 QUIZ

1. Is the following statement correct? Noise signals are always undesirable.

2. Partition noise is not a problem in

    A. vacuum tubes.

    B. bipolar transistors.

    C. MOSFETs

3. Is Johnson's noise an example of IF noise?

    A. No

    B. Yes

4. Which of the following statements is correct?

    A. The greater the bandwidth the greater the system noise.

    B. Noise and bandwidth are not related.

5. Flicker noise is produced by all amplifying devices. It is an example of

    A. white noise.

    B. pink noise.

6. How does AVC/AGC cause noise in a receiver?

    A. It injects noise into the RF and IF circuits.

    B. When the receiver input signal is weak it raises the gain of the receiver.

7. Is it possible to inject noise into a receiver just by connecting an antenna to the antenna terminals?

    A. No

    B. Yes

8. Is flicker noise present in an FET device?

9. Will increasing the bandwidth of an amplifier increase its noise?

10. Is it possible to increase the noise in a system by using the wrong kind of replacement resistor?

11. Is the following statement true? Most of the noise that is generated inside a system is introduced at the output end.

12. Which is noisier, a pentode tube or a triode tube?

13. Before tracking noise or intermittents, disconnect or defeat all _____ circuits.

14. As a general rule, which of the following is best as a DC source for defeating a closed-loop circuit?

    A. Rectifier supply

    B. Battery pack

15. Random motion of electrons in a material is called _____.

16. The name of the logic gate that can change any waveform to a sine wave is called a (an) _____.

17. On a JK flip flop when there is a clock transition from 1 to 0 the flip-flop _____.

18. When the JK flip-flop is in its resting or normal condition, the inputs to J, K, and the clock are all logic _____.

CHAPTER NINE

Servicing Digital Logic and Microprocessor Equipment

Overview

In many ways troubleshooting digital logic and microprocessor circuits is the same as troubleshooting analog circuits. For example, signal injection, signal tracing, symptom analysis and diagnostics are all useful for tracking down troubles in a digital or microprocessor circuit.

The main difference is in the types of signals that you are dealing with and the types of test equipment that are used. There are only two levels of signal voltage to be dealt with: logic 1 and logic 0. In most of the logic systems, logic 1 is the positive power supply voltage. In today's systems this usually means +12V, +3V, +3.6V or +5V. However, there are some specialized circuits where other voltages are used. Logic 0 is almost always zero volts.

When you first learned about linear systems, you had to learn some basic things in order to get started. For example, you had to study and learn about the following:

Mathematics

Power sources

Basic components

Combination of Basic Gates (Amplifiers, oscillators, and other circuits)

Types of signals

Basic test equipment

Test procedures

To work in digital systems, the same range of basics is necessary. An understanding of basics is the key to efficient troubleshooting in all types of systems.

This chapter reviews some of the basics of digital systems, including the microprocessor. Many of the same fundamental concepts for digital circuits that are used for linear systems have been followed in this book.

OBJECTIVES

What do multiplexers do?

What are A/D and D/A converters?

What does a microprocessor do?

POWER SOURCES AND PROPAGATION DELAY

Three major logic families are in use today. They are similar in regard to their use of basic components, called gates, but differ in other ways. For example, the power supply requirements are different for each, see *Table 9-1*. There are CMOS systems that are designed especially for operation at +5V. They are used extensively in microprocessor systems.

In addition to the power supply voltage requirements, there is a difference in propagation delay for each family. *Propagation delay* is the time required for the output of a logic component to reflect a change that has occurred at the input. The ECL family is the fastest, and the CMOS family is the slowest. However, you will not be required to measure those times.

The difference in voltages and times explains why you can't substitute a component from one family into a system made with another family. For example, there are CMOS logic integrated circuits that have the same pinouts as corresponding TTL ICs, but you cannot substitute one for the other.

	Family		
Voltage	TTL	CMOS	ECL
Logic 0	0V	0V	0V
Logic 1	+5V*	4.5-12V†	-5V

TTL - Transistor-Transistor Logic
CMOS - Complementary Metal Oxide Semiconductor
ECL - Emitter-Coupled Logic

* - Regulated
† - May not be regulated

Table 9-1.

THREE-STATE DEVICES

For three-state devices, also called tristate devices, see *Figure 9-1*. These circuits have the ability to open or close a line that carries logic pulses.

The circuit in *Figure 9-1a* is called a *three-state buffer*. It passes the signal, without modification, when the disable terminal is at a logic 0 level. When a logic 1 is delivered to the disable terminal, the input signal is prevented from passing through. Also, the output terminal goes to a high impedance (open-circuit) condition.

The circuit in *Figure 9-1b* has an active low input on the disable terminal. So it is necessary to deliver a logic 0 to the disable terminal to prevent the signal from passing through and to put the output in a high impedance state.

The circuit in *Figure 9-1c* inverts the data when the disable terminal is at logic 0.

An example of the use of tristate logic is shown in *Figure 9-2*. The data output lines from the integrated circuit pass easily when the disable terminal is at logic 0.

If data is introduced to the line by circuit X, the integrated circuit could be destroyed if a logic 0 and logic 1 appeared on the line at the same time. However, the three-state devices can be used to isolate the IC from the lines while circuit X is in operation. In some cases it will be necessary also to have tristate devices at the output of circuit X.

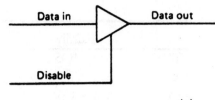

(a)

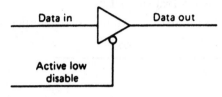

(b)

Figure 9-1. Examples of three-state devices.

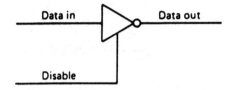

(c)

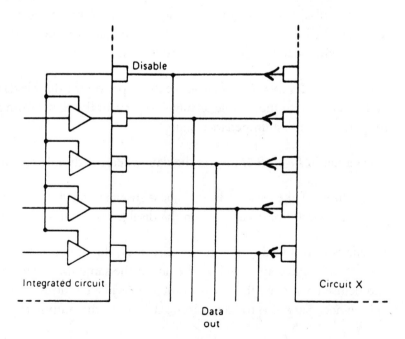

Figure 9-2. The tristate devices protect the integrated circuit when circuit X is delivering data out.

Enables, like three-state devices, permit a logic signal to be either passed through or blocked. The output depends on the enable input signal and the choice of AND or NAND gate.

In the circuits shown in *Figure 9-3*, an open switch prevents the input signal from passing through. The output will always be logic 0 if it is made with an AND gate. It will always be 1 if it is made with an NAND gate. A closed switch enables the circuit. Observe that the output is inverted for the NAND compared to the AND circuit.

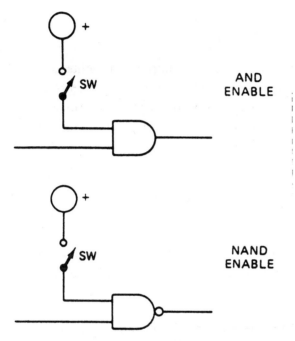

Figure 9-3. Two enable circuits.

Following are definitions of important terms.

Registers are short memories. They are used to hold a digital number for a certain period of time. Registers are used extensively in internal microprocessor circuitry as well as in digital circuits. There are four possible registers, as indicated in *Figure 9-4*.

Frequency counters and *frequency dividers* are used extensively in electronic circuits. They are usually made with flip-flops. Frequency counters count *up* in binary numbers. Programmable counters can be set, using a binary code, to count to a predetermined value and then reset. Frequency dividers count *down*. They are used to divide an input frequency to a lower value, and like counters, they use flip-flops. For an application of frequency dividers, refer to the discussion on phase-locked loops in Chapter 7.

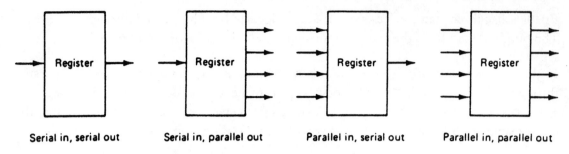

Figure 9-4. The four kinds of registers.

Flip-flops are sometimes called bistable circuits. The output is at either of two levels — usually referred to as high and low, or Q and NOT Q. Switching from one output to the other may involve the use of a clock signal and dynamic circuit, or it may be done with a static combination of input signals. If a clock signal is delivered to a toggled flip-flop, the output is one-half the input frequency. In counters and dividers many flip-flops may be connected internally.

Data flip-flops are used extensively in static memories (without a clock signal) such as *static RAMs (Random Access Memories)*.

Logic comparators are used to compare two logic signals. They produce an output only when the inputs are identical. The simplest logic comparator is the exclusive NOR, which was discussed earlier in this book.

Decoders take a logic level input and converts it into a specific output. A good example is the seven-segment decoder used in a basic counting circuit, as shown in *Figure 9-5*. Here the input is a binary count, and the output controls the segments in such a way that a decimal number is displayed to represent the binary input.

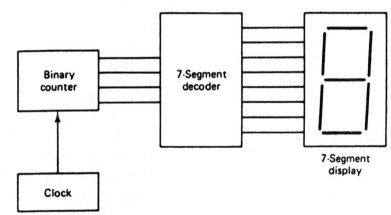

Figure 9-5. Use of a seven-segment decoder to change a binary number to a decimal number.

236 Electronic Troubleshooting and Servicing Techniques

Encoders are the opposite of decoders. They are used to convert some input into a logic count or a logic number. A good example of encoders is in the keyboard electronics that converts a keystroke into a binary number that can be understood by a microprocessor. Each key on the keyboard is *encoded* with a different binary number so that the microprocessor can distinguish between keys being operated in the system.

Displays are used as readouts for a logic system. Seven-segment displays like the one in *Figure 9-5* are very popular, but they have only a limited number of alphabet letters that can be displayed. Alphanumeric (numbers and letters) symbols are usually delivered to a dot-matrix combination like the one shown in *Figure 9-6*. Those displays may be either LEDs (light-emitting diodes) or they may be made with liquid crystals that change the amount of reflected light according to an input voltage. Another popular display is the gaseous type, which utilizes gas ionization to produce light. As a general rule, you will find that gaseous displays require much higher voltage than either liquid crystal or LED displays.

Interfaces are used when it is necessary to convert from one voltage level to another. An example is in a gaseous display. The logic output of a microprocessor might be at a 5-volt level, which would be insufficient to light most gas displays. Therefore, some kind of an interface is utilized so the 5 volts is actually used to control the higher DC voltage. A popular interface is the *optical coupler* of the type shown in *Figure 9-7*. The input signal produces a light from the LED which turns on an output device such as a phototransistor or photo SCR (light-activated SCR, or laser). The two circuit voltages are isolated from each other because only a light source is used for triggering the output device.

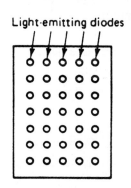

Figure 9-6. A dot matrix that uses light-emitting diodes for displaying alphanumeric characters.

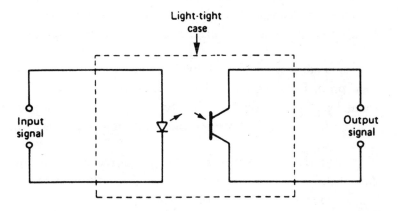

Figure 9-7. One type of optical coupler.

Servicing Digital Logic and Microprocessor Equipment

Knowledge as a Valuable Troubleshooting Aid

As in analog circuits, knowledge of circuit theory and circuit behavior is one of the most important troubleshooting aids. Certainly, you cannot expect to troubleshoot an enable circuit if you don't know how an AND gate works. That idea is carried through to all of the digital circuitry that requires troubleshooting and servicing.

You wouldn't be reading a book on troubleshooting if you didn't know how electronic circuits work. This is especially true of digital logic circuits and microprocessor circuits. However, the brief review of some of the basic components and circuits given in previous chapters will be useful because knowledge is, after all, a troubleshooting tool.

Test Equipment Useful in Troubleshooting Logic and Microprocessor Circuits

The first step in troubleshooting any electronic system, after a thorough visual inspection, is to measure the power supply voltage.

In logic and microprocessor systems, an accurate voltmeter should be used because the power supply in logic circuits and microprocessor circuits is very often stiffly regulated. Because of the close tolerance of the power supply output, a digital voltmeter is highly desirable for that measurement.

As a rule, voltmeters should *not* be used for measuring logic levels at various points in a system. One reason is because it would be necessary to make a decision between logic 1 and logic 0 in many cases. For example, in a 5V logic system, is 2.8V satisfactory for a logic 1?

To eliminate these decisions, logic probes are much more useful. Another advantage of a logic probe is that you don't have to look up from the work each time you make a measurement. Usually the probe has some type of light indicator near the top so that you can see the level of the logic signal being probed without moving your head.

That second reason may look like begging for an excuse to use a logic probe. However, after two or three hundred measurements with a voltmeter you would very quickly get the idea that a probe is much more convenient.

Another advantage of using the logic probe is that it is possible to catch a very short-duration glitch. A glitch is an undesired spike voltage in a logic circuit. The inertia of a

VOM and the time constant of a DMM are usually so high that they do not register the short-duration glitches. However, the glitches are very important because they can cause trouble in a digital or microprocessor system. It may be the very trouble that you are looking for. The glitch would be missed if a logic probe was not used.

Some oscilloscopes, especially those having bandwidths of 20 megahertz or less, will miss the very short-duration glitches also. So, in this rare case, the logic probe is actually better than an oscilloscope for some types of measurements.

When using a logic probe, always be sure to start by probing the DC supply voltage. That should give you a logic 1, but it is possible to have glitches on the power supply output. They will be indicated by the pulse indicator on the probe.

Logic Pulsers

A logic pulser does the same job in digital systems as a signal generator does in analog systems. It injects a desired signal for the purpose of testing a component or circuit or system. A combination of a logic pulser and logic probe is sometimes sold at a reduced price.

Figure 9-8 shows a troubleshooting problem where a combination pulser and probe can save time by preventing a wrong interpretation of a measurement. In *Figure 9-8a*, the logic probe shows a logic 0 output of the AND gate. However, it cannot be determined if this is due to a defective AND gate or if the circuit has been grounded at this point.

Note

The circuit should have a logic 1 output. This can be determined from the AND truth table. If the circuit is grounded at the output, the AND gate will likely be destroyed. However, if you didn't know about the ground you might replace the AND gate and immediately destroy the new one.

Figure 9-8b shows how to test for this type of problem. The pulser injects the signal and the logic probe is used to look for that signal. If the problem is with the gate, the probe will show a pulse signal. If the point is grounded, the probe will indicate no signal.

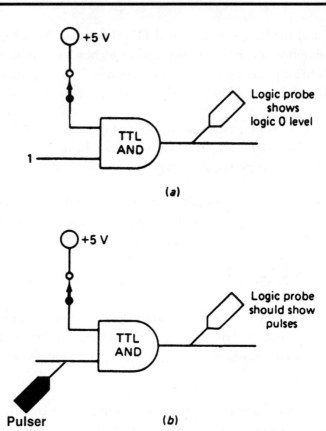

Figure 9-8. How a logic pulser can be used to determine if a gate is not working properly.

FREQUENCY COUNTERS

A frequency counter is a very valuable asset in troubleshooting logic circuits. With that instrument you can locate clock signals and verify their frequency. Of course, the same thing could be done with a good oscilloscope. However, this usually requires you the use of a basic calculation. After tracking the clock signal through several divide circuits, those calculations become tiresome.

It should be mentioned here that some oscilloscopes have frequency counters built in. They are very useful for this type of measurement because you can count the frequency and you can also look at the waveform. However, this type of equipment is expensive.

There is also the additional complexity that must be taken into consideration if the test equipment is down for any reason. In other words, the cost of repairing the equipment in terms of money and labor will be greater than for individual test instruments.

A special type of oscilloscope, called the *logic analyzer*, is also useful for troubleshooting. It is a multitrace oscilloscope that displays as many as six or eight (or more) logic signals simultaneously. That, in turn, allows them to be compared for coincidence and timing of pulses. As with other specialty oscilloscopes, analyzers are more expensive than ordinary oscilloscopes. Their added expense must be weighed against their potential use and the type of troubleshooting they are needed for.

Microprocessors

A good way to think of a microprocessor is to consider it to be a memory operator. That concept can also be applied to computers. All of the useful information in a microprocessor system is stored in a memory. The job of the microprocessor is to store, retrieve and process information in memory.

Since the microprocessor and computer do the same type of work, this discussion on troubleshooting will be directed toward the microprocessor system. One vast difference between the two systems is in the amount of memory involved. Computers use microprocessors to control a very large amount of memory.

It is very important to know that much of the troubleshooting required for these systems centers around mechanical problems. Plugs and sockets are especially troublesome. The pins become loose and sometimes broken. A loose pin can cause impulse noise and intermittent circuit operation, which are difficult troubleshooting problems. The methods used to locate the sources of these problems are not much different from the methods for analog circuits.

The unit connected through the plug will often show symptoms. For example, if the defective plug connects a memory section, that memory will likely be unable to store and return information (at least on the line with the defective pin). Since only one of the many pins is likely to be a problem, scope each output, one at a time, while putting a varying physical pressure on the plug. Look for a disturbance on the scope screen.

An Example of a Microprocessor System

An example of a microprocessor system is shown in *Figure 9-9*. This system requires a very stiffly regulated +5V supply. The clock signal is needed for timing all of the operations in the system. Without this clock signal, *nothing* will happen in the system.

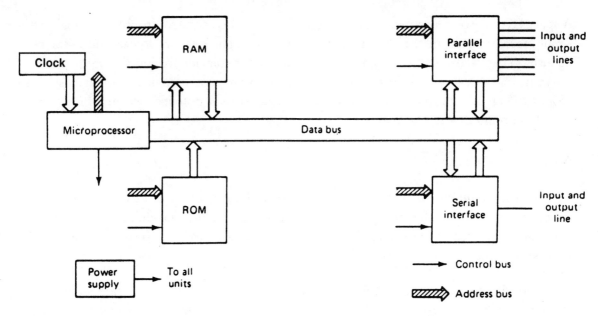

Figure 9-9. Basic microprocessor system.

ROM (Read Only Memory) stores the programs. The programs in ROM are usually put in by the manufacturer. However, there is equipment to program ROMs in the field.

RAM (Random Access Memory) is used by the microprocessor for storing and retrieving numbers. The numbers represent various operations of the system.

The *interface* protects the system from overloads and undesirable external signals.

A *bus* is a combination of wires. The bus may have 8 wires for data and 16 wires for addresses. The addresses are used for locating information in the memories.

The *control bus* carries control signals for operating the various sections. For example, it can set the interface so that information can be delivered to the microprocessor from the outside world. The control signals can also be used to set the interface so that the microprocessor can deliver information to the outside world.

Figure 9-10 shows how a microprocessor is used in a simple four-function calculator. In this case, the microprocessor system contains all of the basic sections shown in *Figure 9-9*. However, they are all in a single integrated circuit. This application is for a dedicated microprocessor.

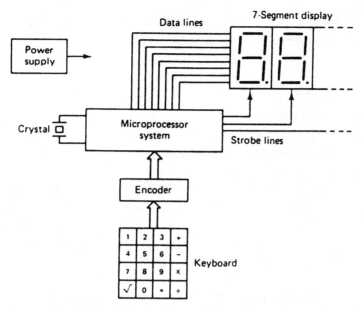

Figure 9-10. Use of a microprocessor in a calculator.

A *dedicated microprocessor* is one designed and programmed to do a specific job. In this application, that job would be performing four functions (add, subtract, multiply and divide). This microprocessor cannot be programmed to another task. this is the main difference between the microprocessor and the dedicated microprocessor.

The power supply (in this case a battery) is a good starting point for troubleshooting. The crystal in this particular application is external to the microprocessor system, but in some modern systems the crystal is located internally. The advantage of locating it externally is that crystals sometimes crack and need to be replaced.

Note

In modern electronics any system that costs less than $100 is often considered to be non-repairable. The reason is that the labor for fixing it can be greater than the cost of the system.

Starting with the keyboard we will discuss the simple addition of 2 + 3. The microprocessor continually polls the keyboard. In other words, it continually looks for an input signal from the keyboard.

- The first step in the addition problem is to enter the number 2. When you push the 2 key on the keyboard the microprocessor takes the binary equivalent of that number 2 (which comes out of the encoder) and puts it into a temporary memory inside the microprocessor. Some manufacturers call the temporary memory the *register*, others call it the *accumulator*.

- The next keystroke is the plus sign. When the plus key is pushed, the encoder sends a coded number to the microprocessor. That number tells it to go to ROM and fetch a program for adding two numbers. Think of the programs being a plan for performing a certain job. That program is entered into a special section in the ROM where it is stored by the microprocessor for future use.

- The third keystroke is for the number 3. The binary equivalent of that number is delivered through the encoder to the microprocessor where it is temporarily stored. In this case, the number is stored in the second accumulator, or another register.

- When you push the equal sign key, a coded message from the encoder tells the microprocessor to start the process of addition using the program that was fetched from ROM. The plan for adding the numbers is sometimes referred to as an algorithm. It is the procedure that must be followed to get the answer.

- Microprocessors have an internal section called the ALU (Arithmetic Logic Unit). The ALU is called on to actually add the two numbers and store the answer in RAM. From there the microprocessor sends the output data through data lines to the seven-segment display.

- It may be necessary to use buffers (not shown) on the data lines in order to drive the display. This would be especially true for a gaseous or LED display. Liquid crystal displays do not require a high voltage or current for their operation, so the microprocessor may be able to deliver signals directly to the display.

- The data lines connect to all of the individual sections, each being part of a seven-segment display. The microprocessor sends a code for the first digit and the strobe line enables the first digit. So the first digit is displayed.

- For example, suppose the first digit in the answer is a number 4. Then the microprocessor delivers power for the proper segments to get the digit 4, the strobe output would enable the first digit so it could be displayed. The power for the second number is then sent through the data lines. (Remember, the data lines connect to all of the digits in the display.) Now the second strobe line enables the second digit and it is displayed. This process continues for as many digits as there are in the display.

The numbers are lighted in sequence so rapidly that your eye believes they are on at all times. By lighting the digits one at a time, the power supply drain is greatly reduced. This is the advantage of strobing the display.

TROUBLESHOOTING A MICROPROCESSOR SYSTEM

Disregard the fact that a calculator is probably a non-repairable device because a calculator of this type can be replaced for a few dollars. Suppose you need to repair it, and you have completed the preliminary checks. The next step is to isolate the problem.

Assume the problem is that the second digit in the display does not light. This could be because the strobe voltage is not being delivered or because the second digit is defective. If the digits can be replaced singly, the next step is to disconnect the strobe line and see if you can light the problem digit with a separate power supply. If you can light it, the trouble is in the microprocessor and it will have to be replaced. If you can't get the display to light with a separate supply, it must be replaced. This is usually a very difficult, if not impossible, job. Only a few types of strobed displays have replaceable elements.

Troubleshooting is a matter of thinking logically by asking yourself: What could be the cause of this trouble? Then your knowledge of the system enables you to determine what procedure is necessary to test for that cause.

Microprocessors are electronic large-scale integrated circuits that are used to control memory. In a microprocessor system or computer, the microprocessor is able to utilize all of the information stored in very high-digit memories. The microprocessor provides special codes for getting the information into and out of memory, operating upon the coded information and delivering the result to some output destination.

Microprocessors are *dynamic*, so there is always a clock signal associated with them. Because of the wide variety of inputs and outputs to microprocessor systems, it is often a good idea to resort to diagnostics for troubleshooting these systems. This assumes, of course, that the preliminary tests do not show a faulty circuit or component. It also assumes that you have tested the mechanical parts when they are a possible problem.

Summary

Many of the troubleshooting procedures for analog systems are the same as for a digital system. The test equipment is different. The logic probe is preferred over a voltmeter and logic pulsers are used instead of signal generators. Nevertheless, they are used in a similar way for signal tracing and signal injection.

As with linear systems, your best troubleshooting aid is your knowledge of how digital and microprocessor systems work.

In computer systems the best way to locate a defective section is by diagnostics. However, remember that plugs and mechanically-operated parts are prime suspects for troubles in computer systems. Check those first.

If you have not already done so, take time to memorize the symbols for the gates and the truth tables.

Chapter 9 Quiz

1. Which of the following best describes the general purpose of a microprocessor?

    A. It implements memory.

    B. It is the basic part of a calculator.

2. An oscilloscope with a relatively narrow bandwidth, such as 20 MHz, is likely to miss a

    A. glitch.

    B. voltage level.

3. What are the only logic levels used in digital and microprocessor systems?

    A. 2 and 1

    B. 1 and 0

4. A microprocessor that is designed to do a specific job and cannot be programmed is called a

   A. dedicated microprocessor.

   B. special microprocessor.

5. Which of the logic families will *not* work with a +5V power supply?

   A. CMOS

   B. TTL

   C. ECL

6. To convert a keystroke into a binary code you need

   A. a decoder.

   B. an encoder.

7. Identify the gate that is sometimes called a comparator.

   A. Exclusive OR

   B. Exclusive NOR

8. Identify the circuit that is sometimes called a bistable circuit.

   A. NOR gate

   B. Flip-flop

9. A clock signal is delivered to a toggled flip-flop. The output frequency is

   A. one-half the clock frequency.

   B. twice the clock frequency.

10. A combination of wires that conveys a parallel logic signal (called a word) is a

    A. logic train.

    B. bus

11. Name four things a logic probe can identify.

12. A signal generator for digital circuits is called a _____.

CHAPTER TEN

Soldering Techniques for Repair and Replacement

Overview

Despite the possibility that it may be viewed as *negative teaching* this chapter has much to say about what you should NOT do when replacing and soldering parts. Some of this material is taken from bulletins issued by the National Aeronautics and Space Administration (NASA). Other sources are publications supplied to technicians by manufacturers of consumer electronic equipment.

When troubleshooting a system, knowledge and test equipment are utilized to find the cause of the trouble. When replacing a component you are, in effect, putting your signature on the work done. If you do unprofessional work you are saying that you don't care enough about your craft to learn how to do it right.

Doing a professional job is not just a matter of pride in your work. It can save you time (and sometimes money) by not having to do the work over again. As a general rule, if you follow the best practices your work will not require rework.

OBJECTIVES

The most common fault in professional soldering.

How a cold solder joint can be recognized.

Special techniques required for surface-mount technology.

Why strong mechanical connections are not desirable for soldering.

Characteristics of a good soldered connection.

SOLDERING TECHNIQUES

How would you classify soldering? Is it an art? Is it a technology? Both terms can be used to describe soldering. The American Heritage Dictionary gives several definitions of art. One is "a craft or trade and its methods." Another is "any practical skill." The same dictionary defines technology as "the application of science especially in industry or commerce." Soldering fits all of these descriptions and more.

THE SOLDER

Soldering is a popular method of connecting circuits. One reason for its extensive use is the low cost of the solder connection. Another reason is its high reliability (when it is done properly).

Most of the solder used today is of the 60/40 type, 60% tin and 40% lead. A somewhat better solder (because it melts at a lower temperature) is the 63/37 variety, which has 63% tin and 37% lead.

Most combinations of tin and lead become plastic before they actually melt. So they have three states: solid, plastic (mushy) and liquid. 60/40 solder is the plastic state, but there is a very small difference between the liquid and solid states.

The 63/37 solder is eutectic at a temperature of 370°. This means that it goes directly from the solid to the liquid state without having a plastic intermediate step. This is one of the reasons that 63/37 is often used for replacing parts on newer systems where the components are very small and sensitive to heat. You do not want to have to hold the solder on long enough to go through a plastic state between solid and liquid. *Figure 10-1* shows the difference between the 60/40 and 63/37 solders when heated.

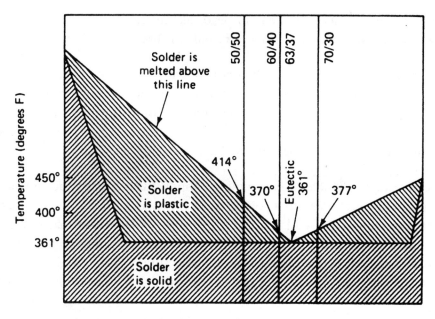

Figure 10-1. Popular solders and their melting temperatures.

Instead of soldering with a combination of lead and tin there are conductive plastic solders. They come in a soft state and are hardened by heat or by adding a chemical called a catalyst or hardener. In either case, the plastic solder is applied first and then hardened.

Plastic solders form a very solid, strong bond, and they are easy to apply. The one that uses a catalyst for hardening is especially useful in surface-mount configurations, where heat must be used sparingly to protect the component.

Some solders (such as silver solder) employ indium. Indium is one of the basic elements. One type of solder combines indium, tin, lead, cadmium and gallium. These solders melt at very low temperatures, but they do not flow easily like the more conventional solders. Some solders melt at so low a temperature that they can be melted with the flame from a match.

Warning

Cadmium is a highly poisonous material! *Never* inhale any smoke from soldering. In fact, a small fan should be part of your bench equipment, to pull the smoke away from you while you are soldering.

Surface-mount components do not employ leads like the resistors and capacitors have, see *Figure 10-2*. Instead, they have a metal interface that is used for soldering.

Figure 10-3 compares a surface-mount component with a conventional component. There are some things you should know about working with surface-mount technology. First, you must *never handle the components*. Second, if you remove a component from a circuit for testing purposes, you should *never replace that component in the circuit*. Use a new one instead. Otherwise, you are going to be liable for unprofitable callbacks.

One of the best ways to handle surface-mount components is to use a vacuum parts holder. It holds the parts by suction so that you do not have to touch them. In some cases these components can be soldered. Also, there is a special solder cream that you can use for soldering onto circuit boards. *Never use 60/40 solder for soldering surface-mount components*. The amount of temperature required to melt the solder is surely enough to destroy the component. As a technician, you should know that solder with an acid flux is *never* used in electronics.

The most convenient solder is the type that has the proper flux (usually resin) in the center or core of the solder. With that type, the flux is applied every time you solder. The flux is needed for cleaning the surfaces in order to make a better solder connection.

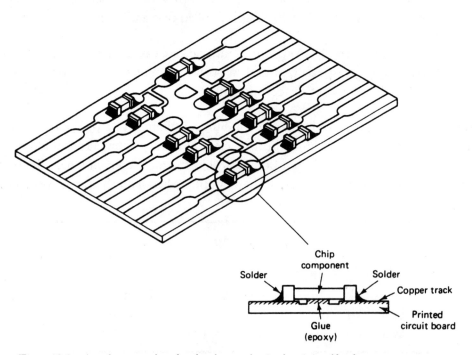

Figure 10-2. A surface-mount board and a close-up showing how it is soldered.

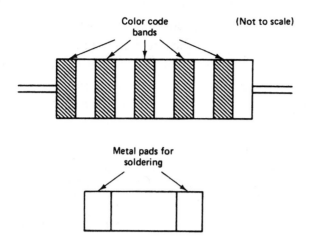

Figure 10-3. Comparison of a conventional resistor with a surface-mount resistor.

SOME IMPORTANT DONT'S

Before discussing soldering techniques, it is a good idea to review some of the things that you must *not* do when making a soldering connection. Some technicians get impatient while they are waiting for the solder to cool and harden. They develop a bad habit of blowing on the solder to cool it. This should be avoided for the following reasons:

> It causes the solder to cool from the outside in.

> The moisture from your breath is trapped in the solder, which can cause problems.

> It has been shown that blowing on solder can cause high-impedance joints at very high frequencies. You do not want to have to stop and think what frequency the circuit is going to be used on before you decide whether to blow on it or not. The best way is to avoid the practice altogether.

There is another bad practice technicians get into. For some unknown reason they begin to wiggle the solder while it is cooling. This a very bad practice because it can produce spaces between the surfaces being soldered. In all cases, it is undesirable to have empty space around the conductor or between the conductor and the solder.

Another highly undesirable practice that seems to be handed down by word of mouth is to make a very strong mechanical connection before soldering begins. The most important reason for not making the mechanical connection is that it does not serve any purpose. At one time it was thought that a strong mechanical connection was required for making a good reliable electrical contact. Today it is known not to be true.

Mechanical connections make it almost impossible to unsolder the lead or the component once the soldering job has been completed. You may be the one who has to unsolder it in the future when you are troubleshooting the same system. You don't want to make a hard job for yourself, and you certainly don't want to make a hard job and an unprofessional job for another technician. *Figure 10-4* shows how to make the connection and how *not* to make the connection.

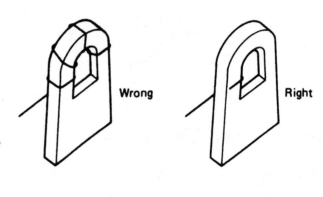

Figure 10-4. Comparison of wrong and right ways to prepare for soldering.

SOLDERING

Figure 10-5 shows terminals ready for soldering. The illustrations should give you an idea of why the mechanical connection is not required.

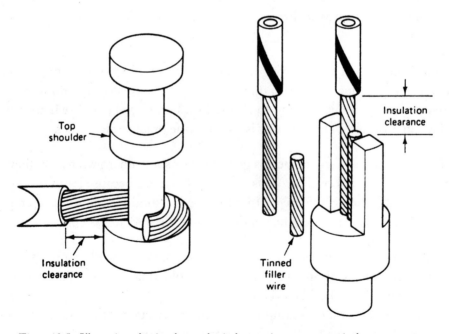

Figure 10-5. Illustrations showing that mechanical connections are not required.

The proper procedure for soldering is shown in *Figure 10-6*. Studying the illustrations will not make you good at soldering, but it will at least give you an idea of how to start. You can read ten books on soldering and you still won't be able to solder until you pick up an iron and some solder and start making connections.

Be very critical of the soldering you are doing. Practice a sufficient amount of time until you can make a good soldering joint. The most common errors in soldering are to use too much solder or too much heat.

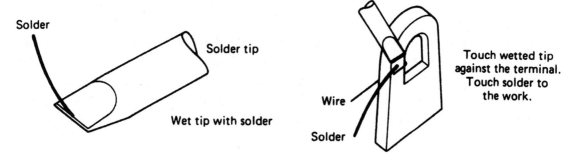

Figure 10-6. The basic soldering procedure.

Too much solder can lead to later troubleshooting problems. Soldering bridges between printed circuit connectors, like the one shown in *Figure 10-7*, often results from using too much solder. That, in turn, results in dropping blobs of solder when you are making the solder connection.

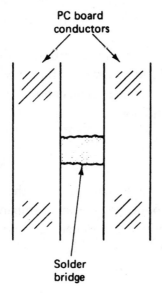

Figure 10-7. A solder bridge.

Soldering Techniques for Repair and Replacement 255

Too much heat can destroy the component that is being soldered. In some cases, if too much heat is used it can destroy the printed circuit board itself. Replacing a printed circuit board is nearly impossible on large complex electronic systems. So, avoid that necessity by being aware of how much heat is being applied and where it is being applied.

Another common soldering problem is to applying too little heat. This causes a dull gray solder connection; it should be shiny. This type of connection is called a *cold solder joint*. It is unreliable and often has a high resistance. In fact, the resistance may be so high that it acts as if no connection has been made. Reheating a cold solder joint usually converts it into one that is acceptable.

Under no condition should you ever use solid wire for making electronic circuits. Solid wire is used for making breadboard circuits, and has a tendency to *work-harden*. If you bend it back and forth several times, it becomes very hard and brittle and will break off easily. Stranded wire, on the other hand, is flexible and much easier to work with. A joint made with stranded wire will be more reliable because it does not work-harden like solid wire.

Inspecting the Solder Connection

As mentioned before, be very critical when inspecting a soldering connection. Practice until it can be done right. If the solder connection is made properly, the solder will be shiny rather than a dull gray.

Another characteristic of a good solder connection is that you should be able to see the strands and the outline of the wire when the solder has cooled. If you can't tell where the wire ends and where the terminal begins, you have used too much solder.

It is not uncommon for some of the flux to remain on the surface of the solder after it is cool. Always take a cotton tip soaked in alcohol and remove that flux. It won't hurt anything except it might gather dust particles and make the terminal unsightly. Also, it is not professional to leave the soldered terminal in that condition.

The Soldering Iron

One of the problems that occurs during soldering is that the tip of the soldering iron becomes dirty after you have made a number of connections. Some companies make a soldering iron holder with a built-in sponge. They advise you to dampen (not soak) the sponge and wipe the soldering tip on it frequently to make sure that the tip is clean. If you don't do this, your solder connection will have bits of dirt in it.

Two kinds of soldering tips are available. One is a solid metal tip. The manufacturer says it is OK to file those tips occasionally to remove oxidation allowing for a good, clean, heated iron when you make your soldering connections.

The other kind of tip is plated. This type is usually recognized because the plating is shiny, especially when it is new. Never file this kind of tip. Doing so will remove the plating and a new tip will be needed.

Unsoldering Connections

A new component cannot be installed until the old one is removed. The trick is to get the component out without destroying the circuit board and the terminals. Thus, great care must be taken when unsoldering. One of the best techniques is to use a solder sucker. It is a rubberized flexible bulb that sucks up the solder through a tube. Another type of solder sucker uses a spring-loaded piston. By using the solder sucker you can remove all of the melted solder. Then the component or the connection will be easy to remove.

Instead of a solder sucker, many technicians prefer a soldering wick. It is a braided wire that removes the solder by a capillary-type action. The procedure is shown in *Figure 10-8*. Simply insert the end of the wick into the melted solder and the solder runs up the braid. Then cut the end off and the wick is ready for use again. The soldering wick allows for area-specific control when unsodlering. You can even shape the end of the braided wire to remove solder from small places. The wick should be about the same width as the conductor from which you are removing the solder. Do not use a soldering wick that is too wide or too narrow. Some technicians keep several sizes on hand.

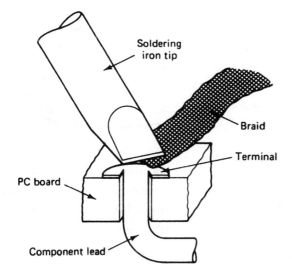

Figure 10-8. Using a wick to remove solder.

Soldering Techniques for Repair and Replacement

After you have removed the solder and the component or wire, the next step is to make sure the terminal is clean before replacing the wire or component. This may require additional heating with the iron. Again, be careful that you do not overheat and destroy the terminal, board or components.

ADDITIONAL SOLDERING TIPS

Before you apply the soldering iron to the connection to be soldered, be sure to *tin* the iron. This simply means to melt solder over its surface. The reason for doing that is so the melted solder will form a good interface. It gets the heat to the place where you want it when you are soldering. Also, make sure the component leads or wire are tinned so that the solder is already fused with the metal. This way you do not have to hold the iron onto the surface for a long time in order to get the solder flowing and to complete your connection.

Technicians who do professional work prefer to have a soldering iron with a controlled temperature. They are more expensive, but they permit you to set the temperature to a fixed value. It is especially important when soldering very small components.

The type of soldering gun shown in *Figure 10-9* should *never* be used for soldering modern electronic circuits. It is nearly impossible to control the heat in this type of iron, even with the soldering guns that have the two-step trigger for two levels of heat.

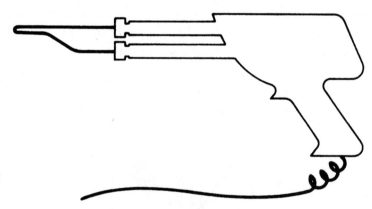

Figure 10-9. This type of soldering gun is not recommended.

Another reason for not using this type of soldering gun is that the secondary of this gun (which is the tip) is insulated from the primary and from other conducting paths. Thus, static charges can build up on those tips that can readily destroy a MOSFET or other field effect device. *Figure 10-10* shows the schematic of the soldering gun.

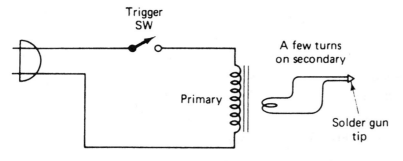

Figure 10-10. This is the circuit for the soldering gun in Figure 10-9. Note that the secondary is isolated so that electrostatic charges can accumulate and destroy MOSFETs.

No matter what kind of iron you use, always make sure that it does not have static charges on it, especially if you are working on electronic systems.

When you are soldering heat-sensitive components, components that can be easily destroyed by heat, it is a good idea to use a heat sync of the type shown in *Figure 10-11*. The theory is that the heat will flow up the lead and into the heat sync rather than into the component.

Although this next point is not directly related to soldering, it is a problem that should be addressed as far as transistors and other semiconductors are concerned. It has been shown that when cutting pliers are used on the leads of a transistor or other semiconductor device, it is possible for a shock wave to travel up the conductor and destroy the component internally. Avoid this by using a finger as a shock absorber between the cutters and the component.

Much has been said about the destruction of MOSFETs by handling. The destruction is due to static electricity. Many of the newer MOSFETs and integrated circuit MOS devices are buffered in such a way that static voltages are routed around the components, so the problem of static electricity destruction has been greatly reduced. Nevertheless, you should employ reasonable precautions when working with those devices. Keep the leads together before they are soldered and make sure that the tip of your iron is grounded.

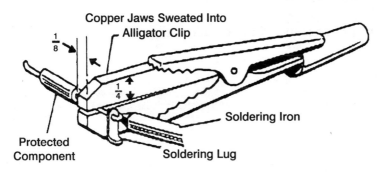

Figure 10-11. A heat sink.

SUMMARY

The knowledge of everything from the characteristics of a good solder connection to the types of solders and soldering tools, combined with the practice of the technology, can prepare you to make soldering a worthy personal signature.

CHAPTER 10 QUIZ

1. When soldering in electronic circuits,

    A. always use acid flux to clean the surfaces.

    B. never use acid flux to clean the surfaces.

2. If a solder goes directly from the solid state to the liquid state, it is said to be

    A. eutectic.

    B. electric.

3. Is the following statement correct? After soldering a wire to a terminal, you should be able to see the outline of the wire and strands of the wire.

4. Which of the following statements is correct?

    A. Never file the tip of a soldering iron.

    B. It is OK to file the tips of some types of soldering irons.

5. While waiting for a soldered connection to cool, you should

    A. not get into the habit of blowing on it.

    B. blow gently on it to hasten the cooling.

6. Instead of a solder sucker, some technicians prefer to use

    A. stranded wire.

    B. braid.

7. Is the following statement correct? While waiting for solder to cool, you should wiggle the conductor. That way you can tell when the solder has hardened.

8. The fumes from molten solder are highly poisonous if the solder contains

    A. tin.

    B. cadmium.

9. How do you remove flux that has not burned off during the soldering process?

10. Name two ways that plastic solders are hardened.

11. Which types of components should never be reconnected into the circuit after they have been removed?

12. If a solder goes directly from the solid to liquid state when heated, it is said to be _____.

13. Which type of flux is *never* used to solder electronic components and circuit?

14. Should a strong mechanical connection be made when soldering a conductor to a terminal?

15. Two common mistakes in soldering are too much _____ and too much _____.

16. What is the cause of a cold solder joint?

17. To remove flux from a solder connection after it has cooled, use a cotton swab and _____.

18. Which type of soldering tip should *never* be filed?

19. What is a solder wick made from?

CHAPTER ELEVEN

Low-Cost Homemade Testing Devices

Overview

Troubleshooting ability depends directly on the abilsity to use manufactured test equipment. There are, however, some basic test instruments that can be built. Some of these instruments have been mentioned in previous chapters, but not identified as homemade.

When is it a good idea for students and technicians to construct test equipment? Here are a few examples:

> When the test equipment cannot be purchased because it isn't being manufactured. One reason it isn't being manufactured is that it is so basic that it cannot be priced profitably.

> When the test equipment is needed for certain specialized tests.

> When less expensive test equipment is the only alternative until more funds are available.

> Because it is interesting to make (and try) electronic projects.

> For the learning experience.

The examples chosen for this chapter are intended to illustrate constructed test equipment. You can find many other examples in magazines and books. It doesn't take much time to build the test equipment in this chapter.

You may be disappointed to find that construction is not described down to the finest detail. That allows you to use the things you have on hand rather than purchase parts and supplies. Also, much of the fun in building a project is in laying out the work.

Despite the many positive ideas about building test equipment, it is still a good idea to use manufactured equipment whenever possible. This is especially true if you are working in a position where your activities are being observed by people who depend on, and accept, your professional expertise; for example, owners of expensive equipment that you are servicing. Those customers will not be impressed (usually) by the fact that you made the test equipment yourself.

OBJECTIVES

Battery pack (used for bias and for defeating closed loops)

Signal injector

Demodulator probe

Resistor substitution box

In-situ transistor tester

Adapter for measuring AC current

NONPOLARIZED ELECTROLYTIC CAPACITORS

Because electrolytic capacitors are polarized, they can only be used in DC and pulsating DC circuits. However, if you are making your own projects, or if you have some special application, it is important to have a very large capacitance that can be used in an AC circuit.

Figure 11-1 shows how a nonpolarized capacitor can be made. The diodes (Da and Db) are called *steering diodes*. The broken arrows show the electron current flow in this circuit when A is negative with respect to B. In other words, if B is common, this is the negative half cycle. The electrons flow through Da and then into and out of Ca during this half cycle.

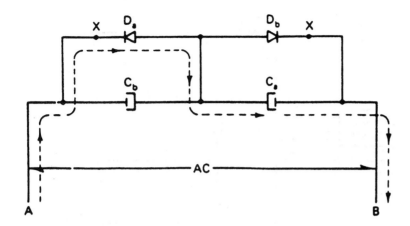

Figure 11-1. A nonpolarized electrolytic capacitor.

When the voltage reverses so that A is positive with respect to B, the electron current will flow through Db, and then into and out of Cb.

Make sure the diodes can handle the current for the application you have. Remember that the *charging current* for an electrolytic capacitor (like Ca and Cb) is very high and can be in the order of hundreds of amperes. For that reason a surge-limiting resistor in series with the diodes may be needed. If so, add a series resistance of less than 100 ohms at the points marked with an X in the schematic. Also, it is a good idea to have a fuse in series with terminal A or B to protect the components in case of heavy surges from the AC source.

Use heavy-duty rectifier diodes with a high current rating.

Voltages to Check Calibration

If you are using a triggered sweep scope, you are sure to have an internal calibrator that permits you to read voltages of waveforms on the display. Likewise, if you are working in electronics, you surely have a multimeter that can measure voltages. However, in both cases, it is a good idea to check the calibration of your instruments periodically. *Figure 11-2* shows a basic zener diode circuit with three voltages for calibration.

The circuit is very basic in design. Follow the design procedure given here for making any open-loop regulated supply. For example, if you want a 9V source to replace or substitute for 9V batteries in equipment, you can use the design procedure that will be given.

All of the zener diodes in this circuit should have the same current rating. So, if they have identical voltage ratings, it follows that the power rating (which is usually given) should be the same for each zener.

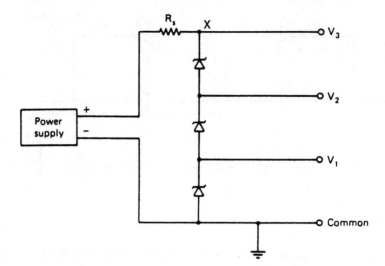

Figure 11-2. A simple voltage calibrator.

The power supply voltage should be at least 50% higher than the highest voltage at point V_3. So, if the zener diodes are rated at 5V each, then the voltage at output terminal V_3 will be 15 volts. The power supply voltage should be the value that is calculated as follows:

Zener (total) voltage x 1.5 = Input power supply voltage

If zener diode total voltage is 15V:

Power Supply Volts = 15V x 1.5 = 22.5V

Now you know the power supply voltage and you know the voltage at V_3. So, the voltage across R_S is the difference between those two voltages, or 7-1/2 volts. If you disregard the current flow in the external load, then the resistance of R_S can be calculated by Ohm's Law. Here is the calculation:

$$R = 7.5V/I_T$$

As a rule, you can disregard any current flow at the terminal because you are using this to calibrate high-impedance oscilloscopes and meters. If, however, there is current flow in the external circuit, that current must be added to the zener diode current flowing through R_S.

If you have a bench power supply, set it to the voltage input value calculated and try the circuit. Then package it in a convenient box with external terminals. Don't forget that checking the calibration of your scope and voltmeter can easily be done with a dry cell. The 1.5 terminal voltage of a dry cell remains the same even though its internal resistance is high due to old age.

Of course, if you try to draw current from an old dry cell, the voltage drop across the internal resistance will subtract from the 1.5 volts. This would give you an unreliable terminal voltage. Remember that oscilloscopes and voltmeters have very high input impedance and they draw very little current. So you can usually disregard the internal voltage drop in the cell.

As a rule, if the calibration of 1-1/2 volts is correct, all of the calibration of the instrument can be relied upon. The circuit of *Figure 11-2* gives better choices.

USING A VOM FOR A HIGH-VOLTAGE MEASUREMENT

Figure 11-3 shows an example of a series-resistance voltage divider. It can be used to drop a high voltage to a lower value that can be measured with your VOM. The proportional method of calculating the voltage across R_5 is shown in *Figure 11-3*.

Note that the voltage divider shown in the illustration divides the voltage to one-tenth its total value. Therefore, measuring across R_5 on the 0 to 500V range is an easy matter. You simply multiply the voltage measurement by 10 to get the actual input voltage. Here it is presumed that 5000V is the absolute maximum voltage, so it is safe to use the 500V scale on the meter.

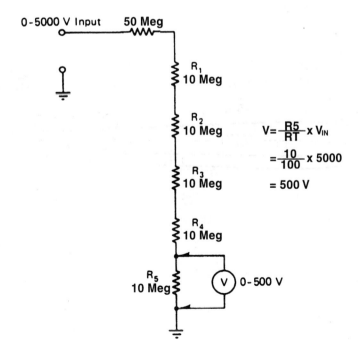

Figure 11-3. *A voltage divider that permits measurement of high voltages.*

This kind of voltage divider is very easy to design. It is basically the same as the voltage divider as used in high-voltage probes for measuring the second anode voltage of CRTs in television receivers and the cathode voltage in the oscilloscope. (In an oscilloscope, it is a common practice to hold the second anode voltage at common, or near common, and operate the CRT with a high negative voltage on the cathode.)

A Basic Logic Probe

Figure 11-4 shows a logic probe that is easy to construct from a discarded pen. An LED is used as an indicator and a series resistor limits the current flowing through the LED. In this particular case, it is assumed that a logic probe will be used in circuits where a logic 1 is +5 volts and a logic 0 is 0 volts.

You can make this probe compact and carry it with you, but remember that it is always better to use a professional logic probe when one is available. The homemade probe is only used for a quick troubleshooting procedure.

Professional logic probes can tell you whether there is a logic 0 present or an open circuit, something that is not possible with the simple probe in *Figure 11-4*. When the LED doesn't light, you don't know if the circuit is open or if you are probing a place where the circuit is grounded. Also, the probe in *Figure 11-4* cannot tell you if a glitch or pulse is present. A professional logic probe can do those things.

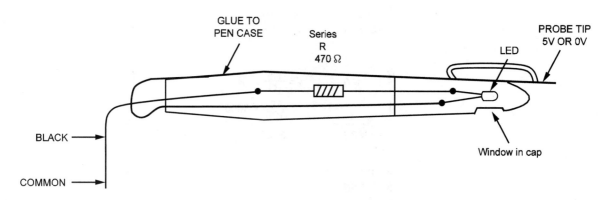

Figure 11-4. A very basic logic probe.

A Basic Noise Generator for Use in Signal Tracing

You will remember that in signal tracing a signal generator of some type is used at the input. The procedure is to follow the signal through the circuitry (using an oscilloscope or voltmeter). In this case, it will be assumed that you are using an oscilloscope.

The circuit of *Figure 11-5* shows a basic noise generator. Note that the silicon diode is reverse biased so that you are producing a small leakage current in the reverse direction. Incidentally, germanium diodes work very well in this application. The reverse current through the diode produces a noise output.

The output test leads are delivered to the antenna terminals of the high-frequency receiver. Then, using an oscilloscope, you follow the noise signals through the RF and detector sections. As soon as you come to the point where you can't see the noise signal, you have passed the cause of the system being inoperative.

When you first connect the circuit to the antenna terminals, adjust the variable resistor for maximum noise at the output of the first RF amplifier. The receiver should be tuned to a frequency where there is no station.

The 300-ohm and 75-ohm terminals are most common for television and radio receivers. Note that there is a switch position without a resistor. The resistors are only needed if there is an isolation capacitor inside the receiver in the antenna line.

Instead of using an ordinary silicon diode, you may be able to obtain a noise diode, which is specifically made for the purpose of generating noise. That, of course, will give you better results.

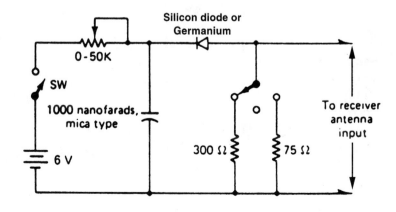

Figure 11-5. An example of a noise generator.

Low-Cost Homemade Testing Devices

Measuring AC Current with a VOM

Most voltmeters and VOMs are unable to measure an AC current, especially a relatively high current.

The setup in *Figure 11-6* is convenient when you have to make the AC current measurement. It consists of two terminals for connecting the AC lead to the power plug and two pin jacks that you can use for your VOM leads. There is a 1-ohm and 10-ohm resistor in series with the AC current being measured.

Remember that if a 1-ohm resistor is used, then the voltmeter will directly display the current flow. If you use the 10-ohm position, you must divide the voltmeter reading by 10 to get the value of current.

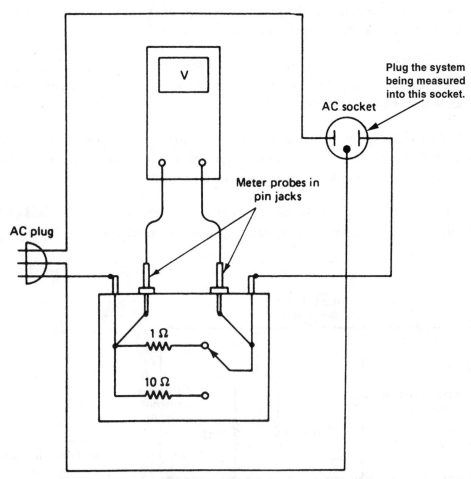

Figure 11-6. A method of measuring AC current.

It is very important that you use wire-wound high-power resistors and allow plenty of room for ventilation if you construct this device. The output socket is used as a receptacle for the system in which you are measuring the current.

Everything in the circuit of *Figure 11-6* is constructed except the meter and meter probes. With the finished instrument, insert the AC plug into the power line and plug the system (for which you are measuring the current) into the socket. Set the switch and connect the meter, then measure the AC current. Always start with the voltmeter on the highest range.

A Resistor Substitution Device

You are probably familiar with the use of decade boxes. They are useful for building new circuits and for troubleshooting.

Decade boxes are made with both resistors and capacitors. The resistor type enables you to select a desired value of resistance or to vary that resistance over a range of values in order to get an ideal ohmic value for substitution into the circuit.

Figure 11-7 shows an alternative to the resistor decade box. In this case, variable resistors are used. In a manufactured decade box there is a separate resistor for each step. The schematic shows the connections for three variable resistors in a box. The cutaway view shows a suggested way of laying out the top of a box.

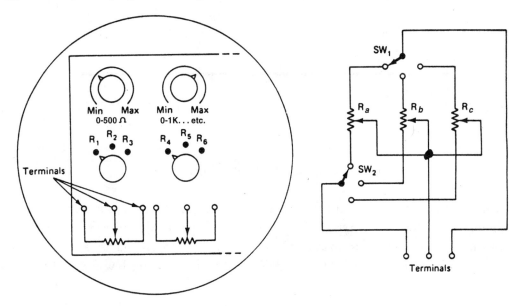

Figure 11-7. A resistance substitution box.

Low-Cost Homemade Testing Devices 271

The variable resistor in the cutaway view is an illustration printed on the box. The symbol shows the user how the terminals are connected internally. *Table 11-1* shows the values of the variable resistors in the schematic as well as for two more ranges. By properly connecting the terminals, you can get just about any combination of resistance value that you want.

One thing should be made clear: A variable resistor should *not* be operated with the arm close to either of the ends. If that happens you should go to the next step. The reason for this rule is that it sometimes places high currents through a relatively small value of resistance and the potentiometer is unable to dissipate the heat.

Resistor	Range of resistance
R_a	0-500
R_b	0-1K
R_c	0-5K
R_d	0-10K
R_e	0-50K
R_f	0-100K
R_g	0-250K
R_h	0-500K
R_i	0-1 Meg
R_j	0-2 Meg
R_k	0-5 Meg
R_l	0-10 Meg

Table 11-1.

DETECTOR PROBE

Figure 11-8 shows a basic schematic for a detector probe that you can easily make. The diode (D) should be a point contact type. The point contact types made today are actually a form of hot carrier diode. They are made with an interface of metal against a semiconductor material. The advantage of this kind of diode is that it has a very low forward breakover voltage. Therefore, there will be an output signal at the coaxial cable even though the input signal strength is very low.

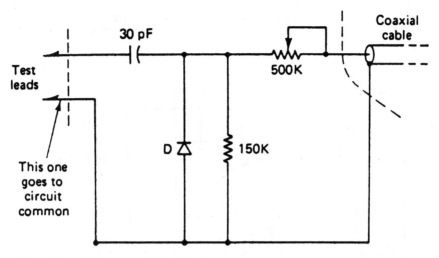

Figure 11-8. A detector (demodulator) probe.

One way to build this probe is to use as a small box to enclose the parts that are between the broken lines. That way you can use test leads without having a cumbersome probe. The best way to build the circuit would be to construct it in some kind of small case so that the circuit is very close to the end of the test leads. This minimizes distributed capacitance.

If you probe in a circuit that has a modulated signal and the coaxial cable goes to an oscilloscope, you should get a display of the modulation only. If the output goes to a voltmeter, then you will see a deflection that is proportional to the average value of the modulation signal. In either case, you will be able to probe the output even though the RF frequency is too high for the instrument you are using.

Figure 11-9 shows an example of a circuit that produces a rectangular output. This circuit can be used for signal injection in circuits over a wide range of frequencies. The harmonic content of the signal is very high, so it works for both audio and RF signals.

The signal is produced by a multivibrator. The output signal is taken from the collector of one of the multivibrator transistors. The variable resistor permits you to adjust the frequency over a range of values. This is not the only way to make a signal injector. Some technicians prefer to use a basic 555 integrated circuit timer. It can be operated as an astable multivibrator, and it will also produce a pulsed output. You might try a 555 circuit as an astable multivibrator instead of the one in *Figure 11-9*.

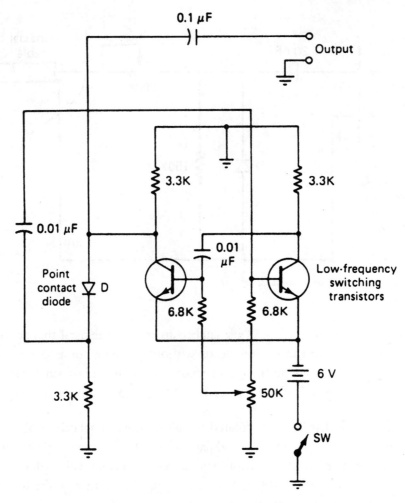

Figure 11-9. An example of a signal injector (any pulse or wave generator will do).

A Simple In-Situ (On-Site) Transistor Tester

This circuit was suggested by Mike Olszewski Smith of West Palm Beach, Florida. It has been tested for both voltage amplifier and power amplifier transistors.

In-situ testers can be used for checking transistors out of a circuit and checking them while wired into a circuit. The theory of this tester is that the transistor, when used in an oscillator circuit, must be good if it supports oscillation. Remember that an oscillator is just an amplifier with regenerative feedback. So, when this device checks a transistor, it is used as the amplifier.

Figure 11-10 shows an example of a simple in-situ tester that you can easily build. In this circuit the oscillator is in a Hartley configuration. The primary of the output transformer is used to shift the phase for feedback, and the 0.047 microfarad capacitor is used to set the audio frequency.

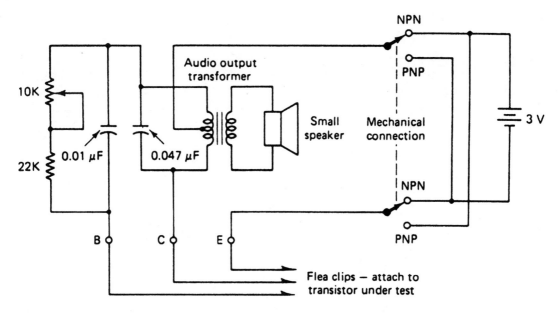

Figure 11-10. An in-situ tester.

The feedback circuit is through the 10K and 22K resistors. The capacitor across those resistors ensures that high-frequency transients will not produce regenerative feedback and cause the amplifier to oscillate at a high frequency.

Both NPN and PNP transistors can be tested. An audio sound from the speaker indicates that the transistor under test is good. If you are checking a transistor while it is in the circuit, be sure that the power supply for that circuit is disabled. Usually this is accomplished simply by turning off the system.

Low-Cost Homemade Testing Devices

APPENDIX A

Experimenter's Logic Probe

The logic probe shown below is good for practicing troubleshooting techniques. However, for your work in logic systems you will want one that is more elaborate. You will find them advertised in Radio Electronics and other trade magazines. Mount the LED as close to the probe tip as you can get it. I have seen several *homemade* jobs, and the one I liked best was mounted in a small pill box.

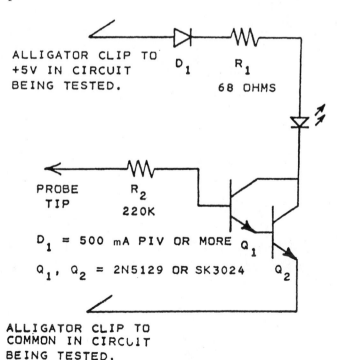

APPENDIX B

Binary Counting

0	000000		21	010101
1	000001		22	010110
2	000010		23	010111
3	000011		24	011000
4	000100		25	011001
5	000101		26	011010
6	000110		27	011011
7	000111		28	011100
8	001000		29	011101
9	001001		30	011110
10	001010		31	011111
11	001011		32	100000
12	001100		33	100001
13	001101		34	100010
14	001110		35	100011
15	001111		36	100100
16	010000		37	100101
17	010001		38	100110
18	010010		39	100111
19	010011		40	101000
20	010100		41	101001
			42	101010

APPENDIX C

Boolean Algebra

$A \cdot A = A$ A and A equals A

$A \cdot 0 = 0$ A and 0 equals 0

$A \cdot 1 = A$ A and 1 equals A

$A \cdot \overline{A} = 0$ A and NOT A equals 0

$\overline{\overline{A}} = A$ NOT NOT A equals A

$A + 0 = A$ A OR 0 equals A

$A + A = A$ A OR A equals A

$A + 1 = 1$ A OR 1 equals 1

$A + \overline{A} = 1$ A OR NOT A equals 1

$A + \overline{A}B = A + B$ A OR NOT A AND B equals A OR B

$\overline{A} + AB = \overline{A} + B$ NOT A OR A AND B equals NOT A OR B

Repairing and Replacing

APPENDIX D

Quiz Answers

Chapter 1

1. Quantitative

2. Qualitative

3. No. With the switch open, the voltmeter should display the applied voltage.

4. Lower. Higher current, flowing through the transistor, increases the drop across the internal resistance of the supply. That, in turn, lowers the output voltage of the supply.

5. Negative

6. AC and Reverse

7. No. The 1.5V supplied by the ohmmeter is too low to be effective in testing for leakage.

8. A

9. Power Supply

10. An electrolytic filter capacitor

11. A. A leaky electrolytic capacitor can cause too much current to flow through the surge-limiting resistor.

The surge-limiting resistor (R_1) protects the diode during the first few cycles that occur when the capacitors are charging. The protection is needed when the circuit is first energized.

If the capacitors have an excessive leakage current, the resistor will burn out. Remember, it may be designed to act like a fuse.

Always check the electrolytic capacitors before replacing the surge-limiting resistor!

12. A. If the diode permits a reverse current flow there will be an AC voltage across the electrolytic capacitors. An AC voltage across an electrolytic capacitor will destroy it.

13. B. If an analog meter is poorly damped you might be able to see vibration of its pointer. However, it is doubtful. In that case you wouldn't be measuring the DC voltage across the capacitor.

The best way to determine the condition of an electrolytic capacitor is to use an ESR meter. The initials ESR stand for Equivalent Series Resistance. It is the combination of series resistance and parallel leakage resistance of the capacitor.

An ESR meter is very useful for determining the condition of an electrolytic capacitor. It can also be used for other capacitor types.

14. A. Reversing the diode will reverse the voltage across the electrolytic capacitors. That, in turn, will quickly destroy the capacitors.

15. C. You can use the square wave test for a quick overview of the amplifier's frequency response. A good frequency to use for checking an audio amplifier is 2500 hertz.

The rise time equation given in this chapter is empirical. In other words, it is based on experience rather than being derived mathematically. Although the rise time measurement is quantitative, there are more accurate ways to determine the bandwidth of an amplifier. Those ways are discussed in a later chapter.

16. B. Remember that the power supply voltage must be checked under load. In other words, check the voltage while the equipment is energized.

Note: Don't confuse the terms load and load resistance. Load is the amount of current delivered by the source of voltage. Load resistance is the amount of opposition to that current.

17. B. The more likely cause of the low voltage is leakage current in C_1, C_2, or a defective (shorted) component in the power supply load.

If the voltage goes to its normal value when the output capacitor (C_2) load circuit is disconnected, then the trouble is in that circuit. If the output remains low, then C_2 is a likely cause.

Resistors are not as likely to change value compared to the chance that there is a leakage current through a capacitor.

18. B. The filter capacitors charge to the peak voltage. Remember that AC voltages are stated as RMS values unless otherwise stated.

    Peak voltage = 100 x 1.414   (1.414 is the square root of 2)

    Peak voltage = 141.4V

There will be a voltage drop across the diode, surge-limiting resistor, and filter resistor, so the output will be somewhat lower than the calculated peak value.

19. A. Assuming the meter is a Simpson 260 or its equivalent, it deflects to the average value of the input. The average value of a sine wave over a cycle is zero. So the meter displays zero volts. (Even though the pointer on the meter is well damped, you may see some pointer vibration.)

20. Lay it on its back because the bearings in the meter are then in the position for the least amount of friction.

21. No. Never connect an ohmmeter in a circuit with a voltage.

22. Equivalent Series Resistance (ESR)

23. Bandwidth = 0.35/rise time

24. Digital (high impedance) voltmeter or oscilloscope

25. A. Oscilloscopes with bandwidth ratings of 50 MHz or greater are common in logic system troubleshooting.

26. B

27. B

28. B. ECL requires a -5V supply

29. NOT M

30. TTL, CMOS, ECL

31. AND

32. NOR

33. B+ (the power supply output voltage)

34. B

Chapter 2

1. B. Observe that the emitter is more positive than the collector. That is another way of saying that the collector of the PNP transistor is negative with respect to the emitter. The input signal is at the base and the output is at the collector, so, this is a common emitter amplifier.

2. B. Be very careful when working on circuits with enhancement MOSFETs!

3. No. As with bipolar transistors the voltage across the input electrode (emitter or source) is opposite in polarity to the control electrode (base or gate). In other words the voltage across the emitter or source works against the bias.

4. A. Among other things, R_9 acts to pull down the base voltage. If that resistor is open the base voltage will float toward the positive voltage of the supply. There won't be enough voltage drop across R_8 to save the transistor, so, it will be saturated, and probably be destroyed.

5. A. It is the AVC (also called AGC) filter. It prevents signals at the detector from getting into the bias for Q_2.

6. A. The power supply voltage must be positive so that the gate and drain are negative with respect to the source.

7. B. As noted in the answer to #6, the power supply is the most positive point. All other points are positive with respect to common. Since the gate is more positive than the drain, it is more positive than common.

8. C. With long-tail bias point 'X' is very near to ground potential (zero volts). So, shorting point 'X' to ground will not destroy either power transistor.

9. To eliminate crossover distortion.

10. P - channel MOSFET. If you missed this question you should review all of the bias circuits.

11. B

12. N-channel JFET and N-channel depletion MOSFET

13. B

14. The power supply voltage

15. Varactor (sometimes called by its trade name, Varicap)

16. A

17. A

18. A

19. Any of the devices

20. B. Regenerative feedback is used with oscillators

21. Exclusive OR

22. NOR

Quiz Answers   287

Chapter 3

1. A and B. Both germanium and silicon rectifier diodes should have at least a 10 to 1 (front-to-back) resistance ratio. In other words, when the ohmmeter measures the reverse resistance it should be at least 10 times greater than the forward resistance. The test will NOT identify the type of diode. The forward voltage test (0.7V for silicon and about 0.2V for germanium) will also identify the type of diode.

2. B. The emitter-base short-circuit can destroy a transistor.

3. B. It is one of the advantages of a meter with a taut band meter movement.

4. B. This is an amplifier distortion test.

5. B. You need to draw current from the battery when testing its terminal voltage.

6. Audio

7. +2 dB

8. B

9. Low-power ohms

10. 1 ohm.

11. 5 megohms

12. A

13. A

14. R X 1

15. A

16. Direct-coupled amplifiers and Class B amplifiers

17. Sine wave

Chapter 4

1. A

2. Frequency domain.

3. B

4. A

5. A

6. Delayed

7. A

8. 2

9. Zero

10. B

11. B. The emitter-base short-circuit can destroy a transistor.

12. B. It is one of the advantages of a meter with a taut-band meter movement.

13. RS flip-flop

14. Nothing; the flip-flop stays in the same output condition.

15. JK flip-flop

16. B

Chapter 5

1. A

2. Capacitor. Use it to bridge the signal from amplifier A, across amplifier B, to the input of amplifier C. There will be a loss of performance, but if amplifier B is at fault you will get a weak signal output from amplifier C.

3. B

4. B

5. A

6. B

7. A

8. B

9. 1695 kHz

10. Detector

11. White

12. Not true. It is necessary to keep the input to the amplifier at a constant amplitude. Some signal generators are designed to maintain a constant output amplitude. This is a very useful feature.

13. Sweep generator, function generator with VCO input

14. Crowding of the frequency at the upper and lower ends of the frequency range.

15. Frequency

16. Screwdriver

17. 112.1

18. B. The high harmonic content of a pulse makes it useful over a wide range of frequencies.

19. C

20. B. Do not be in a hurry to align tuned circuits in a system. Only in rare cases would alignment be the cause of a system failure. (An exception is the tuned circuits in kit radios.) Think of alignment as a last resort in troubleshooting.

21. Your truth table should show that it is an Esclusive OR circuit.

CHAPTER 6

1. No. Eyewitness accounts can be misleading and a waste of time. Use them as a *possible* aid in troubleshooting.

2. No. There are cases where there is trouble but no observable symptoms.

3. A

4. Measure the power supply voltage.

5. Yes. You should remember that a low power supply voltage may affect the weakest circuit in the system. In the case mentioned here, that would be the audio amplifier.

6. Look for hot spots, burn marks, and places where a voltage arc-over may have occurred.

7. Yes. The weaknesses are sometimes mentioned in service bulletins. You should put them in your notebook for future reference.

8. B

9. A. The rectifier diode normally gets warm (or hot) during its normal operation. Suspect hotter components before those that do not get hot.

10. No. It is best not to take the customer's word as final for any part of the troubleshooting procedure.

Chapter 7

1. Closed-loop circuits and intermittents

2. B

3. B

4. A

5. B

6. B

7. B

8. B

9. A

10. C

11. Ease of replacement.

12. Negative, also called degenerative; positive, also called regenerative.

13. Circuits for controlling voltage circuits for controlling frequency

14. DC

15. Open the loop

16. Analog, or linear; digital, or switching

17. Greater efficiency

18. AVC/AGC

19. Control gain of the RF and IF amplifiers

20. The statement is correct.

21. Zener diode

CHPATER 8

1. No. Noise signals can be used for aligning and testing.

2. C

3. A. Johnson's noise covers a wide spectrum, but, IF noise decreases in amplitude as the frequency increases it is not Johnson's noise. The name *pink noise* is sometimes used to refer to IF noise.

4. A

5. A

6. B

7. B. Because of the antenna resistance.

8. Yes

9. Yes

10. Yes

11. Not true

12. Pentode (because of the increase in partition noise)

13. Closed-loop

14. B

15. Intrinsic current

16. Schmitt trigger

17. Toggles

18. 1

Chapter 9

1. A

2. A

3. B

4. A

5. C

6. B

7. B

8. B

9. A

10. B

11. Logic 0, logic 1, glitch, pulse

12. Pulser.

Chapter 10

1. B. Acid flux is highly corrosive and will destroy electronic parts and connectors.

2. A. An alloy solder is eutectic at only one temperature.

3. Yes. Remember that too much solder is a typical problem with solder connections.

4. B. You can file the tips if they are solid but not if they are plated.

5. A. The best way is to let the solder connection cool without your help.

6. B. It is a matter of opinion whether you use braid or a solder sucker.

7. No. Never move the conductor when the solder is cooling. If you do, you may create pockets between the conductor and solder terminal.

8. B. As a general rule, you should make it a practice to avoid inhaling anything but air.

9. Remove the flux with alcohol and a cotton swab. Do not use any other type of cleaning agent because some of them leave deposits that attract dirt. Also, some cleaning agents attack rubber and plastic parts.

10. Heat, catalyst or hardener (the method depends on the type of plastic solder).

11. Surface-mount

12. Eutectic.

13. Acid flux

14. No

15. Solder, heat

16. Not enough heat during soldering

17. Alcohol

18. A plated tip

19. Braided copper wire

INDEX

Symbols

0 signal 18
1 signal 18
1/f noise 216
555 integrated circuit timer 273
60/40 solder 250
63/37 solder 250

A

Absolute zero 213
AC current 7, 87
AC input 7
AC power line voltage 8
AC voltmeter 7, 87
Accumulator 244
Accuracy 93, 96, 112
Active device 45
Address 242
AFC 164, 199
AGC 43, 44, 53, 164, 186
AGC bias 53
AGC/AVC bias 54
AGC/AVC feedback loop 185
AGC/AVC output 165
Algorithm 244

Alignment 189
Alphanumeric character 118
Alternate 128
ALU 244
AM receiver 64
Ambient temperature 54
Amplifier 14, 198, 200
Amplifier configuration 38
Amplifier distortion 137
Amplifier lissajous test 137
Amplifier noise 217
Amplifier signal 160
Amplifier stage 13
Amplifier voltmeter test 102
Amplifying device 38, 40, 48, 49, 53, 199
Amplifying system 6
Amplitude 10, 23, 125, 133
Analog current meter 5
Analog meter 4, 85, 87, 88
Analog system 4, 17
Analog VOM 83
Analog voltage regulator 193
Analog voltmeter 83
Analyzer 241
AND gate 14, 20, 21, 106
Antenna 64, 187, 219
Arithmetic logic unit 244

Astable multivibrator 273
Attenuator 119
Audio amplifier 64
Audio distortion 177
Audio modulation 161
Audio push-pull amplifier 69
Audio system 220
Audio waveform 64
Auto-ranging digital voltmeter 4
Automatic frequency and phase comparator 203
Automatic frequency control 164, 185, 199
Automatic gain control 164, 186, 188
Automatic volume control 188
Autoranging meter 4
Autotransformer 198
AVC 44, 53
AVC bias 53
AVC line 66
AVC/AGC bias voltage 191
AVC/AGC circuit 186
AVC/AGC DC control voltage 186
AVC/AGC feedback loop 189
Average 131
Average value 131
Avogadro, Amedeo 213
Avogadro's law 213
Avogadro's number 213, 214

B

B 214
Bandwidth 11, 12, 55, 56, 59, 73, 117, 122, 149, 162, 214
Base 40, 45
Base bias 68
Base bias voltage divider circuit 55
Base current 40
Base voltage 102
Base-to-collector capacitance 43
Base-to-collector junction 41
Basic circuit 14, 17
Basic digital device 38
Basic gate 105, 140
Basic logic gate 2
Basic probe 18
Battery 3, 6, 22, 95
Battery bias 53

Battery pack 192, 264
Battery supply 4, 6
Battery terminal voltage 95
Battery-operated equipment 4
Beta test 75
Bias 38, 40, 48, 53, 103, 192
Bias circuit 56
Bias substitution 192
Bias voltage 49
Binary arithmetic 15
Binary count 107
Binary counting 113
Binary number 107
Bipolar amplifier 55
Bipolar device 40
Bipolar power transistor 181
Bipolar transistor 7, 40, 41, 43, 53, 54, 63, 102, 216
Bipolar transistor amplifier 41
Bipolar transistor circuit 49
Bipolar transistor power amplifier 7
Bistable circuit 236
Blanking control 120
Boltzmann's constant 213, 214
Boolean algebra 15, 109, 110
Bounceless switch 140
Brownian motion 211
Buffer 29, 244
Bus 242

C

Calibration 90, 112, 126, 127, 192, 265
Capacitive feedback circuit 198
Capacitor 7, 8, 14, 30, 42, 43, 58
Carrier wave 64
Cathode 40, 43
Cathode ray tube 120
Cathode-to-plate current 43
Center tap 154
Charge carrier 46, 48
Charging current 265
Chop 128
Circuit 14, 15, 18
Circuit board 17
Circuit breaker 2
Circuit resistance 89

CK terminal 147
Class A amplifier 102, 134
Class A operation 48, 61
Class AB operation 61, 68
Class B amplifier 105, 135
Class B operation 49
Class C amplifier 135
Class C operation 49
Clear terminal 148
Clipping 72
Clipping distortion 138
Clock circuit 145, 221
Clock signal 145, 146, 221
Closed-filament 97
Closed-loop circuit 185, 186, 198
Closed-loop feedback circuit 220
Closed-loop system 185, 188, 220
CMOS 27, 29, 30, 87, 141, 232
CMOS family 232
CMOS IC 29
CMOS system 232
Cold solder joint 256
Collector 45, 55, 102
Collector current 40
Collector junction 41
Collector voltage 197
Combination RF amplifier 64
Combinational logic 14, 105, 113
Combined circuit 146
Combined gate 110
Commercial tester 98
Common 50
Common base 50
Common denominator 176
Common emitter 52
Common gate 50
Common source 52
Common-mode connection 60
Communications receiver 190
Communications system 192, 220
Complementary amplifier 63
Complementary metal oxide semiconductor 29
Complementary MOSFET 27
Complementary vacuum tube 63
Complicated impedance-coupling network 13
Component 14, 15

Component test feature 132
Computer 145
Conducting half cycle 49
Conducting region 46
Configuration 50, 52
Constant-current device 60
Control bus 242
Control electrode 40, 43, 48
Control grid 40, 43, 45
Control system 145
Converter 64
Cooling fin 61
Counter 14
Coupled circuit 57
Coupling capacitor 57, 69
Coupling network 58
Current 4, 6, 7, 46, 61, 85, 117, 130
Current measurement 6
Current meter 7
Current regulator 196
Current scale 5
Current-reading meter 90
Curve 118
Cutoff 145, 221

D

Data flip-flop 236
Data line 244
dB scale 84
DC bias voltage 53
DC current 5
DC gain-control voltage 186
DC gate voltage 43
DC level 50
DC measurement 43, 45
DC milliammeter 7
DC path 18
DC polarity 40, 53
DC supply 16
DC voltage 38, 162
DC voltmeter 7
Dead system 157
Decade box 271
Decay time 117
Decimal number 107
Decoder 236

Dedicated microprocessor 243
Defective fuse rating 6
Defective switch 6
Deflection 5
Deflection plate 122
Delay 124
Demodulator probe 94, 264
Demultiplexer 168
Depletion MOSFET 47, 48
Depletion region 43, 46
Detection 64
Detector probe 272
Diagnostic 176, 177, 178, 231
Dielectric 42
Differential amplifier 60
Digital circuit 2, 4, 10, 18, 105, 110, 113
Digital circuit measurement 31
Digital component 110
Digital logic scope 120
Digital meter 4
Digital multimeter 8, 83, 96
Digital probe 83
Digital troubleshooting 2, 14, 18
Digital voltmeter 150
Digital VOM 100
Diode 4, 7
Diode test 101
Dip meter 198
DIP package 17
Direct coupling 59
Direct probe 94
Direct-coupled amplifier 59
Direct-coupled circuit 59
Direct-coupled device 50
Discriminator 165
Discriminator/ratio detector 203
Display 237
Distorted sine wave 136
Distortion 3, 118, 132, 177
DMM 96, 239
Drain 40, 46, 55, 102
Drain current 40, 46
Dry cell 90, 95, 266
Dual flip-flop 30
Dual in-line plastic package 17
Dual-gate MOSFET 44, 45

Dual-trace oscilloscope 122
Dual-trace triggered-sweep oscilloscope 128
DVM 87
Dynamic 245

E

ECL 27, 29
ECL family 232
Elapsed time 125
Electric field 46
Electrode 48
Electrolytic capacitor 7, 8, 99, 100, 264
Electron 18, 45
Electron flow 18
Electron gun 122
Electronic circuit 17
Electronic schematic 21
Ellipse 136, 138
Emitter 40, 45, 50
Emitter-base junction 49, 87
Emitter-base voltage 102
Emitter-collector short circuit 104, 181
Emitter-coupled amplifier 27
Emitter-coupled logic 29
Emitter-resistor bypass capacitor 68, 69
Emitter-to-base short circuit 86
Emitter-to-base voltage 197
Emitter-to-collector connection 75
Emitter-to-collector short circuit 7
Enable 14, 235
Enable circuit 238
Enable input signal 235
Encoder 237
Enhancement MOSFET 40, 41, 48, 54
Enhancement transistor 53
Equivalent circuit 23
Equivalent series resistanc meter 8, 100
ESR measurement 8
ESR meter 100
ESR tester 100
Eutectic 250
Exclusive NOR 236
Exclusive NOR gate 106
Exclusive OR gate 20, 27, 106, 111
Exclusive OR situation 23
External noise 219

External sweep signal 119

F

False bass tone control 69
Family 29
Fan in 21
Fan out 21
Faraday screen 43
Faraday shield 43
Feedback capacitor 69
Feedback circuit 197, 198
Feedback loop 188, 219
Feedback network 199
Feedback system 198
Ferrite iron core 57
Fidelity 189
Field effect 46
Field effect transistor 40, 98, 219
Filament transformer 90
Filament-to-cathode short 97
Filter capacitor 7, 8, 194
Filter resistor 7
Flicker noise 216
Flip-flop 14, 30, 141, 147, 223, 224, 236
Floating ground 192
Flowchart 176, 178
Flux 252
Flyback transformer 196
FM detector 199
FM receiver 190
Follower circuit 50
Form factor 91, 92
Forward bias 54, 62, 74, 196
Forward breakover voltage 272
Forward voltage 102
Forward voltage bias 40
Forward-biased transistor 86
Frequency 10, 11, 66, 117, 125, 126, 163, 185, 197, 199
Frequency calibration 127
Frequency counter 127, 146, 163, 195, 197, 198, 203, 235, 240
Frequency distortion 118, 132
Frequency divider 200, 235
Frequency domain 118, 125, 134, 169, 190
Frequency domain display 125, 161, 162, 165, 190
Frequency domain trace 164
Frequency ratio 137, 149
Frequency response 118, 161, 165, 169
Frequency synthesizer 202
Frequency-determining network 199
Function generator 92, 140, 161, 162
Fuse 2, 4, 5, 7
Fuse rating 6

G

Gain 19, 50, 54, 55, 56, 59, 155
Gallium arsendie 102
Gas constant 214
Gaseous display 237
Gate 20, 21, 24, 27, 30, 40, 46
Gate voltage 46
Gate-to-drain voltage 48
Generator frequency control 133
Germanium diode 102
Glitch 12, 15, 124, 161, 223, 238
Graticule 163
Graticule division 163
Grid 97
Grid-to-plate capacitance 43
Ground 8
Ground potential 46, 52
Grounded cathode 52
Grounded grid 50
Grounded-cathode configuration 50

H

Half cycle 49
Half-wave rectifier 4
Half-wave rectifier circuit 4
Harmonic 124
Harmonic distortion 149
Harmonic frequency 123, 165
Hartley configuration 275
Hearing aid 220
Heat sink 61, 103
Height 9
Heterodyne 67
High voltage arc 2
High-frequency circuit 41

High-frequency distortion 11
High-gain amplifier 13
High-impedance digital voltmeter 13
High-power amplifier 49
High-voltage gain 50
High-voltage probe 94
Higher-level voltage 18
Hole 18
Horizontal amplifier 119
Horizontal oscillator 196
Horizontal sweep 162

I

IC 17
IC logic pinout 146
IC package 222
Identifier 17
IF amplifier 57, 66
IF frequency 156
IF stage 164, 189
IF transformer 69
IGFETS 46
Impedance 87, 233
Impedance coupling 58
Impulse noise 219, 241
In-situ tester 72, 180, 182, 198, 264, 274, 275
Inclusive OR 23
Indium 251
Individual gate 107
Inductance 101
Inductor 14, 58
Industrial schematic 21
Industrial speed control 220
Industrial system 220
Injected signal 163, 164
Input 21
Input signal voltage 61
Insulated gate field effect transistor 46
Insulating material 46
Integrated circuit 10, 16, 18, 20, 27, 29, 76, 140, 146, 222
Intensity modulation 164
Interface 237, 242
Intermittent 13, 185, 209, 220
Intermodulation distortion 138
Intermodulation distortion test 139

Internal logic circuitry 19
Internal modulation 161
Internal resistance 95
Internal sweep signal 134
Intrinsic current 211, 212
Inverse sin 136
Inverter 15, 23
Inverter NOT gate 106
Ion layer 186
Isolation capacitor 95, 157

J

Jewel bearing 93
Jeweled analog meter 93
Jeweled movement 85
JFET 40, 46, 48, 49, 53, 219
JK flip-flop 147, 149, 223
Johnson's noise 215
Jumper 87
Jumper lead 86
Junction capacitance 43
Junction diode 43
Junction field effect transistor 46

K

k 214
Kelvin 213, 214
kT 214

L

Lamp 22
Landmarks 17
Laser 237
Latch 140
LC circuit 199
LCR meter 101
Leakage 8, 100
Leakage current 269
Leakage resistance 8
LED 17, 268
LED display 237, 244
Level shifting 59
Light-activated SCR 237
Linear amplifier 67, 135
Linear circuit 76

Linear closed-loop regulator 195
Linear device 38
Linear distortion 118, 132, 134, 149
Linear system 231
Linear troubleshooting 38
Linear voltage regulator 193
Liquid crystal display 237, 244
Lissajous pattern 135, 136, 137, 140, 149
Lissajous test 136, 138
Load 5, 6
Logic 14, 19, 109
Logic 0 signal 18
Logic 1 output 21
Logic 1 signal 18
Logic analyzer 241
Logic circuit 15, 16, 18, 27, 30, 107
Logic circuit measurement 19
Logic circuitry 19
Logic comparator 236
Logic component 14
Logic family 27, 223
Logic gate 2, 15, 19, 29, 30, 76, 84, 113
Logic IC 16, 30
Logic level 110, 141, 142
Logic probe 2, 18, 146, 238, 268
Logic pulser 239
Logic symbol 149
Logic system 16, 17, 18, 27, 238
Logic timing diagram 120
Logic-level comparator 106
Long-tail bias 52, 60, 63, 72
Low emitter-collector current 75
Low-capacity probe 129
Low-frequency amplifier 57
Low-pass filter 200
Low-power ohm scale 86, 87
Low-voltage power supply 177
Lower-level voltage 18

M

Marker input 164
Marker signal 164
Maximum amplitude 11
Maximum current scale 5
Maximum height 9
Maxterm 15

Mean 131
Measured value 112
Measurement 13, 15
Measurement efficiency 14
Megahertz 11
Memory scope 120
Memory state 142
Metal oxide 46
Metal oxide semiconductor field effect transistor 46
Meter display 5
Meter loading 89
Meter probe 94
Meter resistance 89
Microprocessor 241, 244, 245
Microprocessor system 232, 238, 242
Microseconds 11
MIL symbol 15
Military schematic 21
Milliammeter 1, 7
Minterm 15
Mirror 85
Mirrored scale 112
Mixer 64
Modulated RF signal 70
Modulation 95, 117
Modulation signal 161
MOSFET 46, 49, 55, 63, 219
Motor 61
Motor speed control 202
Multiplexer 166
Multiplier 89
Multitrace oscilloscope 110, 241
Multitrace signal display 110
Multivibrator 145, 273
Multivibrator circuit 145, 221

N

N 214
N region 40
N-channel device 46
N-channel enhancement MOSFET 47
N-channel enhancement MOSFET 48
N-channel JFET 46, 53, 54
N-channel MOSFET 40, 46
N-type material 45

NAND 145
NAND flip flop 141, 143
NAND 20, 23, 106, 141, 143
NAND RS flip-flop 141, 142
NAND SR flip-flop 142
Nanoseconds 15
Negative gate voltage 46
Negative power supply voltage 29
Negative resistance 44
Negative voltage 46
Negative-going bias 192
Network 18
Nicad battery 192
Noise 191, 209, 213, 219, 223, 227
Noise Generator 165
Noise power 210
Noise power equation 214
Nonlinear distortion 138
Nonpolarized electrolytic capacitor 264
Nonsinusoidal voltage 84
Nonsinusoidal waveform 92
NOR gate 20, 26, 106, 141
NOR RS flip-flop 142, 145
Not allowed condition 145, 147, 149
NOT gate 14, 20, 23, 27, 106, 142, 145, 221
NPN bipolar transistor 48, 54
NPN bipolar transistor model 45
NPN transistor 41, 45, 59, 63, 74, 275
Null 87

O

Ohmmeter 8, 38, 73, 75, 90
Ohmmeter scale 96
Ohm's law 266
Op amp logic comparator 203
Open-filament 97
Open-loop configuration 204
Open-loop regulated supply 265
Operational amplifier 38, 60, 71, 72
Optical coupler interface 237
OR gate 14, 20, 22, 106
Oscillations 13
Oscillator 14, 49, 64
Oscillator frequency 157
Oscillator signal 43, 157
Oscillator transformer 69

Oscilloscope 7, 9, 12, 13, 72, 87, 239, 241
Out-of-phase signal 15, 55, 62
Outboard transistor amplifier 76
Output 21, 23, 24
Output current 7
Output logic level 110
Output noise 220
Output signal voltage 61
Output voltage 4, 7

P

P region 40
P-type material 45, 46
Parallax 85, 112, 130
Parallel input 166
Parallel LC 135
Parallel resistance 100
Parallel resistor 7
Parallel transmission 166
Partial short circuit 104
Partition noise 216, 217
Passband circuit 190
Pattern 9
Peak 87
Peak value 127
Peak voltage 163
Peak-to-peak signal 163
Peak-to-peak value 9, 92, 126, 127, 131
Pentode 43, 44, 45
Perfect amplifier 210
Perfect sine wave 136
Period 127
Persistence 124
Phase 185
Phase angle 117, 136
Phase comparator 201, 202, 203
Phase distortion 118, 132, 138
Phase relationship 121
Phase shift distortion 138
Phase splitter 69
Phase-locked loop 185, 200, 202, 235
Photo SCR 237
Pin 17
Pink noise 165
Pinout 30
Plastic solder 251

Plate 43, 102
Plate collector 40
Plate current 44
Plate voltage 44
PN junction 102, 132
PN region 46
PNP transistor 41, 63, 74, 275
Point-by-point plot 149
Point-by-point test 134
Polarity 37, 45, 52, 53
Position control 119
Positive feedback 197
Positive half cycle 49
Positive-going bias 192
Positive/negative bias 52
Powdered iron core 57
Power amplifier 1, 6, 38, 61
Power gain 50
Power MOSFET 48
Power rating 7
Power source 232
Power supply 4, 6, 30, 40, 59, 243
Power supply current 6
Power supply lead 17
Power supply output 7
Power supply output voltage 8
Power supply voltage 4, 10, 30, 51
Power transistor 103
Power-operating VFET 63
Precision resistor 90
Preset terminal 147
Probe 18, 19, 84, 94, 161
Program 242
Programmable counter 235
Programmable counter/divider 203
Propagation delay 15, 29, 232
Public address system 220
Pulse 11, 19, 146, 156, 195
Pulsing signal 18
Pulsing voltage 18
Push-pull circuit 62
Push-pull power amplifier 61

Q

Qualitative judgment 10
Qualitative measurement 2, 9, 31

Qualitative test 2, 149, 165
Quantitative judgment 10
Quantitative measurement 10, 11, 31, 136
Quantitative test 2, 149
Quiescent condition 49, 53
Quiescent point 49

R

R 214
Radar display 121
RAM 120, 242, 244
Random access memory 120, 236, 242
Ratio detector 165
RC network 199
Read only memory 242
Receiver 64, 153
Reception 186
Reception system 64
Rectifier 8
Rectifier diode 265
Recurrent sweep scope 119
Reference voltage 40
Regenerative feedback 73, 197, 275
Register 235, 244
Rregulated power supply 185
Regulated suppl 16
Regulation 196
Relaxation oscillator 195
Relay circuit 23, 26
Reproduction 64
Required value 112
RESET 141
Resin 252
Resistance 8, 12, 75, 85, 86, 87, 88, 95, 99, 100
Resistor 2, 7, 14, 30, 48, 58, 87
Resistor model 48
Resistor noise 210
Resistor substitution box 264
Resistor substitution device 271
Resistor-transistor logic 30
Resting state 142
Restoring force 93, 112
Retrace 120
Reverse bias 41, 43, 46
RF amplifier 43, 57, 64

RF amplifier bias 192
RF circuit 43, 58
RF range 161
RF signal 43, 70, 161
RF transformer 69
Ripple 7, 191, 194
Rise time 11, 12, 117, 124, 149
RMS value 7, 91, 95, 127, 131
ROM 242
Root mean square 131
Round metal package 17
RS flip-flop 140, 142, 149, 223, 226
RTL 27, 30
RTL gate 30
RTL IC 30

S

Saturation 145, 221
Sawtooth generator 119
Sawtooth test 132
Sawtooth time base 135
Sawtooth voltage 164
Sawtooth waveform 119
Scale 4, 5
Scan-derived supply 196
Schematic 18
Schematic symbol 15, 21
Schmitt trigger 149, 150, 226
Scope pattern 9
SCR 237
Screen grid 43
Selection part 64
Self bias 53
Semiconductor 100
Semiconductor circuit 86
Semiconductor device 73
Semiconductor region 42
Semilog paper 133
Serial transmission 166
Series 7, 8
Series capacitor 95, 129
Series resistance 100
Series resistor 94
Series-pass regulator 196
Series-resistance voltage divider 267
SET 141

Seven-segment display 244
Short circuit 8
Short-circuit voltmeter test 104
Sideband 189
Signal 18
Signal current 61
Signal generator 157, 161, 239, 269
Signal injection 14, 17, 154, 155, 156, 157, 160, 169, 231
Signal injector 156, 264, 273
Signal tracer 160
Signal tracing 14, 17, 154, 155, 156, 159, 169, 231
Signal voltage 61
Signal-to-noise ratio 212, 213, 226
Silicon diode 269
Sine wave 48, 49, 95, 136, 149
Sine wave generator 163, 190
Sine wave oscillator 199
Sine wave voltage 91
Single-ended power supply 60
Snow 226
Solder 250
Solder sucker 257
Soldering wick 257
Solid-state equipment 41
Sound distortion 3
Source 40, 46, 50
Speaker 64, 66, 154, 186
Specifications 12
Spectrum 216
Spectrum analyzer 139, 149
Speedometer 85
Spike 15
Spike voltage 12
Square root 131
Square wave 10, 11, 12, 145, 149, 156, 221
Square wave generator 11, 128
Square wave test 10, 11, 132
SR flip-flop 140
Stage 13, 155
Start-up circuit 196
Starting friction 93
Static charge 29, 84
Static RAM 236
Station 64

Station frequency 157
Statistical analysis 3, 180
Statistical approach 3, 30, 199
Statistical method 193
Steering diode 264
Stiction 93
Straight line 136
Strobing 245
Substrate 46
Supercoolant 221
Superheterodyne radio 38
Superheterodyne receiver 190
Supply 4
Supply output voltage 7
Supply regulation 194
Supporting system 198
Suppressor grid 44
Surface mount component 252
Surge-limiting resistor 7
Sweep 127, 163
Sweep analysis 14
Sweep generator 70, 119, 134, 162, 164, 165
Sweep generator test 134
Sweeping frequency 165
Switch 4, 5, 22
Switching regulator 195
Symbol 15, 21, 51, 71, 178
Symptom analysis 3, 176, 231
Symptom 1, 2, 3
Sync amplitude 119
Sync signal 119
Synchronous motor 202
Synthesizing 203
System 14, 15, 154, 157
System noise 227
Systematic approach 17

T

T 214
Tank circuit 135
Taut-band analog meter 93, 112
Taut-band movement 85
Television receiver 190
Temperature 57

Temperature stabilization 68
Temperature stabilization resistor 68
Test 15
Test equipment 3, 12, 17, 264
Test jig 75
Tetrode 43, 44, 45
Thermal agitation noise 215
Thermal noise 215
Thermal runaway 57
Thermocouple meter 130
Three-legged integrated circuit 194
Three-state buffer 233
Three-state device 233
Three-terminal active device 45
Three-terminal device 48, 50
Thyristor 98
Time 126
Time base 122
Time domain 118, 125, 169
Time duration 9
Time interval 117
Time-constant test 99
Timing 118
Timing circuit 17
Timing diagram 110, 111, 113, 120
TO package 17
Toggle 148
Toggle condition 148, 226
Toggled flip-flop 236
Tone control 17
Totem pole 63
Transducer 61, 66, 220
Trnsformer 198
Transformer coupling 57, 58
Transient voltage 161
Transistor 3, 14, 18, 30, 38, 72, 73
Transistor beta 217
Transistor noise 217
Transistor-sensitive amplifier 177
Transistor-transistor logic 29
Triggered-sweep scope 11, 121, 265
Triode tube 43, 45
Tristate device 233
Troubleshooting procedure 3
Truth table 21, 22, 23, 26, 84, 105, 108, 113, 145, 148, 149, 223

TTL 27, 29, 30
TTL family 147, 223
TTL IC 29
Tube 43
Tube amplifying device 104
Tube tester 97
Tuner 14, 64
Tuning capacitor 69
Tuning circuit 43
Twisted flat band 112
Two-input AND gate 21, 106
Two-input exclusive OR gate 111
Two-input NAND gate 27, 106
Two-input NOR gate 27, 106
Two-input OR gate 106

U

Unregulated supply 8
Upside-down PNP transistor 59

V

V-FET 48
Vacuum 43
Vacuum parts holder 252
Vacuum tube 40, 44, 48, 63, 97, 216
Vacuum tube circuit 49
Vacuum tube plate 41
Vacuum tube voltmeter 87
Varactor diode 43
Variable frequency output 11
Variable resistor 17, 48, 49, 193
Variac 90
VCO 201, 202, 203
VCO input 163
Verical amplifier 119, 122
VFET 63
Visual inspection 3
Volt-ohm-milliammeter 12, 267, 270
Voltage 4, 85, 101, 117, 126, 185
Voltage amplifier 66
Voltage arc 178
Voltage decoupling filter 69
Voltage divider 55, 94, 268
Voltage drop 12
Voltage gain 50

Voltage polarity 4
Voltage value 19
Voltage-controlled oscillator 43, 200, 203
Voltmeter 7, 10, 12, 18, 85, 101, 238, 270
Voltmeter measurement 93
Volume control 17, 70, 154
VOM 87, 239, 267

W

Wattage rating 7
Wattmeter 210
Waveform 146
Waveform distortion 118
White noise 165, 215
White noise generator 165
Width 9

X

X-axis 136
X-Y axis 136
X-Y operation 136
X10 sweep expander 11

Y

Y-axis 136

Z

Z-axis 164
Zener diode 29
Zero point 49

ES&T Presents TV Troubleshooting & Repair

Electronic Servicing & Technology Magazine

TV set servicing has never been easy. The service manager, service technician, and electronics hobbyist need timely, insightful information in order to locate the correct service literature, make a quick diagnosis, obtain the correct replacement components, complete the repair, and get the TV back to the owner.

ES&T Presents TV Troubleshooting & Repair presents information that will make it possible for technicians and electronics hobbyists to service TVs faster, more efficiently, and more economically, thus making it more likely that customers will choose not to discard their faulty products, but to have them restored to service by a trained, competent professional.

Originally published in *Electronic Servicing & Technology*, the chapters in this book are articles written by professional technicians, most of whom service TV sets every day.

Video Technology
226 pages - Paperback - 6 x 9"
ISBN: 0-7906-1086-8 - Sams: 61086
$18.95 ($25.95 Canada) - August 1996

ES&T Presents Computer Troubleshooting & Repair

Electronic Servicing & Technology

ES&T is the nation's most popular magazine for professionals who service consumer electronics equipment. PROMPT® Publications, a rising star in the technical publishing business, is combining its publishing expertise with the experience and knowledge of *ES&T's* best writers to produce a new line of troubleshooting and repair books for the electronics market. Compiled from articles and prefaced by the editor in chief, Nils Conrad Persson, these books provide valuable, hands-on information for anyone interested in electronics and product repair.

Computer Troubleshooting & Repair is the second book in the series and features information on repairing Macintosh computers, a CD-ROM primer, and a color monitor. Also included are hard drive troubleshooting and repair tips, computer diagnostic software, networking basics, preventative maintenance for computers, upgrading, and much more.

Computer Technology
288 pages - Paperback - 6 x 9"
ISBN: 0-7906-1087-6 - Sams: 61087
$18.95 ($26.50 Canada) - February 1997

CALL 1-800-428-7267 TODAY FOR THE NAME OF YOUR NEAREST PROMPT PUBLICATIONS DISTRIBUTOR

Semiconductor Cross Reference Book Fourth Edition

Howard W. Sams & Company

This newly revised and updated reference book is the most comprehensive guide to replacement data available for engineers, technicians, and those who work with semiconductors. With more than 490,000 part numbers, type numbers, and other identifying numbers listed, technicians will have no problem locating the replacement or substitution information needed. There is not another book on the market that can rival the breadth and reliability of information available in the fourth edition of the *Semiconductor Cross Reference Book*.

Professional Reference
688 pages - Paperback - 8-1/2 x 11"
ISBN: 0-7906-1080-9 - Sams: 61080
$24.95 ($33.95 Canada) - August 1996

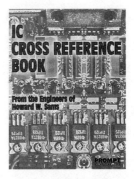

IC Cross Reference Book Second Edition

Howard W. Sams & Company

The engineering staff of Howard W. Sams & Company assembled the *IC Cross Reference Book* to help readers find replacements or substitutions for more than 35,000 ICs and modules. It is an easy-to-use cross reference guide and includes part numbers for the United States, Europe, and the Far East. This reference book was compiled from manufacturers' data and from the analysis of consumer electronics devices for PHOTOFACT® service data, which has been relied upon since 1946 by service technicians worldwide.

Professional Reference
192 pages - Paperback - 8-1/2 x 11"
ISBN: 0-7906-1096-5 - Sams: 61096
$19.95 ($26.99 Canada) - November 1996

CALL 1-800-428-7267 TODAY FOR THE NAME OF YOUR NEAREST PROMPT PUBLICATIONS DISTRIBUTOR

The Howard W. Sams Troubleshooting & Repair Guide to TV
Howard W. Sams & Company

The Howard W. Sams Troubleshooting & Repair Guide to TV is the most complete and up-to-date television repair book available. Included in its more than 300 pages is complete repair information for all makes of TVs, timesaving features that even the pros don't know, comprehensive basic electronics information, and extensive coverage of common TV symptoms.

This repair guide is completely illustrated with useful photos, schematics, graphs, and flowcharts. It covers audio, video, technician safety, test equipment, power supplies, picture-in-picture, and much more. *The Howard W. Sams Troubleshooting & Repair Guide to TV* was written, illustrated, and assembled by the engineers and technicians of Howard W. Sams & Company.

The In-Home VCR Mechanical Repair & Cleaning Guide
Curt Reeder

Like any machine that is used in the home or office, a VCR requires minimal service to keep it functioning well and for a long time. However, a technical or electrical engineering degree is not required to begin regular maintenance on a VCR. *The In-Home VCR Mechanical Repair & Cleaning Guide* shows readers the tricks and secrets of VCR maintenance using just a few small hand tools, such as tweezers and a power screwdriver.

This book is also geared toward entrepreneurs who may consider starting a new VCR service business of their own. The vast information contained in this guide gives a firm foundation on which to create a personal niche in this unique service business. This book is compiled from the most frequent VCR malfunctions Curt Reeder has encountered in the six years he has operated his in-home VCR repair and cleaning service.

Video Technology
384 pages - Paperback - 8-1/2 x 11"
ISBN: 0-7906-1077-9 - Sams: 61077
$29.95 ($39.95 Canada) - June 1996

Video Technology
222 pages - Paperback - 8-3/8 x 10-7/8"
ISBN: 0-7906-1076-0 - Sams: 61076
$19.95 ($26.99 Canada) - April 1996

CALL 1-800-428-7267 TODAY FOR THE NAME OF YOUR NEAREST PROMPT PUBLICATIONS DISTRIBUTOR

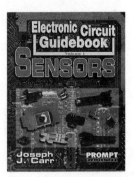

Electronic Circuit Guidebook Volume 1, Sensors
Joseph J. Carr

Most sensors are inherently analog in nature, so their outputs are not usable by the digital computer. Even if the sensor is supposedly a digital output design, it is likely that an inherently analog process is paired with an analog-to-digital converter. In *Electronic Circuit Guidebook, Volume 1: Sensors*, you will find information you need about typical sensors, along with a large amount of information about analog sensor circuitry. Amplifier circuits are especially well covered, along with differential amplifiers, analog signal processing circuits and more. This book is intentionally kept practical in outlook. Some topics covered include electronics signals and noise, measurement, sensors and instruments, instrument design rules, sensor interfaces, analog amplifiers, and sensor resolution improvement techniques.

Electronic Theory
340 pages + Paperback + 7-3/8 x 9-1/4"
ISBN: 0-7906-1098-1 + Sams: 61098
$24.95 + April 1997

Electronic Circuit Guidebook Volume 2, IC Timers
Joseph J. Carr

Timer circuits used to be a lot of trouble to build and tame for several reasons. One major reason was the fact that DC power supply variations would cause a frequency shift or slow drift. Part I of this book is organized to demonstrate the theory of how various timers work. This is done by way of an introduction to resistor-capacitor circuits, and in-depth chapters on various TTL and CMOS digital IC devices. Through simplified equations and detailed graphics, the information presented is perfect for both practicing technicians and enthusiastic hobbyists. Part II presents a variety of different circuits and projects. Some of the circuits include: analog audio frequency meter, one-second timer/flasher, relay and optoisolator drivers, two-phase digital clock and more. *Electronic Circuit Guidebook, Volume 2: IC Timers* will teach you enough that you will not only be able to rework and modify the circuits covered here, but also design a few of your own.

Electronics Technology
240 pages + Paperback + 7-3/8 x 9-1/4"
ISBN: 0-7906-1106-6 + Sams: 61106
$24.95 + August 1997

CALL 1-800-428-7267 TODAY FOR THE NAME OF YOUR NEAREST PROMPT PUBLICATIONS DISTRIBUTOR

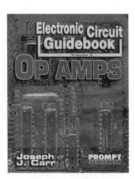

Electronic Circuit Guidebook Volume 3, Op Amps
Joseph J. Carr

The operational amplifier is the most commonly used linear IC amplifier in the world. The range of applications for the op amp is truly awesome – it has become a mainstay of audio, communications, TV, broadcasting, instrumentation, control, and measurement circuits. Third in a series covering electronic instrumentation and circuitry, *Electronic Circuit Guidebook, Volume 3: Op Amps* is design to give you some insight into how practical linear IC amplifiers work in actual real-life circuits. Because of their widespread popularity, operational amplifiers figure heavily in this book, though other types of amplifiers are not overlooked. This book allows you to design and configure your own circuits, and is intended to be a practical workshop aid. Some of the topics covered in detail include linear IC amplifiers, ideal operational amplifiers, instrumentation amplifiers, isolation amplifiers, active analog filter circuits, waveform generators, and many more.

Electronics Technology
273 pages + Paperback + 7-3/8 x 9-1/4"
ISBN: 0-7906-1131-7 + Sams: 61131
$24.95 + August 1997

Electronic Circuit Guidebook Volume 4, Electro-Optics
Joseph J. Carr

Electronic Circuit Guidebook, Volume 4: Electro-Optics is mostly about E-O sensors — those electronic transducers that convert light waves into a proportional voltage, current, or resistance. The coverage of the sensors is wide enough to allow you to understand the physics behind the theory of operation of the device, and also the circuits used to make these sensors into useful devices. This book examines the photoelectric effect, photoconductivity, photovoltaics, and PN junction photodiodes and phototransistors. Also examined is the operation of lenses, mirrors, prisms, and other optical elements keyed to light physics.

Electronic Circuit Guidebook, Volume 4: Electro-Optics is intended to teach the physics and operation of E-O devices, then proceed to circuits and methods for actual application of the devices in real situations.

Electronics Technology
416 pages + Paperback + 7-3/8 x 9-1/4"
ISBN: 0-7906-1132-5 + Sams: 61132
$29.95 + October 1997

CALL 1-800-428-7267 TODAY FOR THE NAME OF YOUR NEAREST PROMPT PUBLICATIONS DISTRIBUTOR

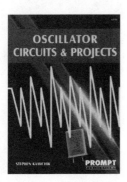

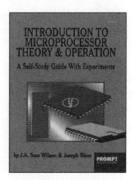

Oscillator Circuits & Projects
Stephen Kamichik

Introduction to Microprocessor Theory & Operation
J.A. Sam Wilson & Joseph Risse

Oscillator circuits usually are not taught at technology and engineering schools and universities. Electronics is a vast field; therefore, some areas of expertise, including oscillators, must be acquired in the field. Oscillator Circuits & Projects helps to make this process easier by presenting the information you need to master oscillator circuitry.

Oscillator Circuit & Projects was written as a textbook and project book for individuals who need to know more about oscillator circuits. Students, technicians, and electronics hobbyists can build and enjoy the informative and entertaining projects described at the end of this book. Compete information about oscillator circuits is presented in an easy-to-follow manner with many illustrations to help guide you through the theory stage to the hands-on stage.

This book takes readers into the heart of computerized equipment and reveals how microprocessors work. By covering digital circuits in addition to microprocessors and providing self-tests and experiments, *Introduction to Microprocessor Theory & Operation* makes it easy to learn microprocessor systems. The text is fully illustrated with circuits, specifications, and pinouts to guide beginners through the ins-and-outs of microprocessors, as well as provide experienced technicians with a valuable reference and refresher tool.

Electronic Projects
256 pages ♦ Paperback ♦ 7-3/8 x 9-1/4"
ISBN: 0-7906-1111-2 ♦ Sams: 61111
$19.95 ♦ April 1997

Electronic Theory
211 pages ♦ Paperback ♦ 6 x 9"
ISBN: 0-7906-1064-7 ♦ Sams: 61064
$16.95 ($22.99 Canada) ♦ February 1995

CALL 1-800-428-7267 TODAY FOR THE NAME OF YOUR NEAREST PROMPT PUBLICATIONS DISTRIBUTOR

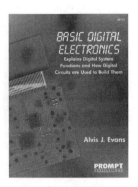

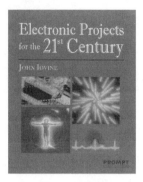

Basic Digital Electronics
Alvis J. Evans

Electronic Projects for the 21st Century
John Iovine

Explains digital system functions and how digital circuits are used to build them! Digital — what does it mean? Why is it that electronic systems are being designed using digital electronic circuits? Find the answer to these questions and more, as you learn the difference between analog and digital systems, the functions required to design digital systems, the circuits used to make decisions, code conversions, data selections, adding and subtracting, interfacing and storage, and the circuits that keep all operations in time and under control.

Learn about logic circuits, flip-flops, registers, multivibrators, counters, 3-state bus drivers, bidirectional line drivers and receivers, and more using easy-to-read, easy-to-understand explanations coupled with detailed illustrations.

If you are an electronics hobbyist with an interest in science, or are fascinated by the technologies of the future, you'll find Electronic Projects for the 21st Century a welcome addition to your electronics library. It's filled with nearly two dozen fun and useful electronics projects designed to let you use and experiment with the latest innovations in science and technology — innovations that will carry you and other electronics enthusiasts well into the 21st century!

Electronic Projects for the 21st Century contains the expert, hands-on guidance and detailed instructions you need to perform experiments that involve genetics, lasers, holography, Kirlian photography, and more. Among the projects are a lie detector, an ELF monitor, air pollution monitor, pinhole camera, laser power supply for holography, synthetic fuel, and an expansion cloud chamber.

Electronic Theory
192 pages ◆ Paperback ◆ 8-1/2 x 11"
ISBN: 0-7906-1118-X ◆ Sams: 61118
$19.95 ◆ April 1997

Electronic Projects
256 pages ◆ Paperback ◆ 7-3/8 x 9-1/4"
ISBN: 0-7906-1103-1 ◆ Sams: 61103
$19.95 ◆ June 1997

CALL 1-800-428-7267 TODAY FOR THE NAME OF YOUR NEAREST PROMPT PUBLICATIONS DISTRIBUTOR

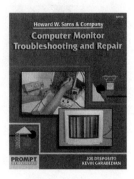

Howard W. Sams Complete VCR Troubleshooting & Repair

Joe Desposito & Kevin Garabedian

 Complete VCR Troubleshooting and Repair contains sound VCR troubleshooting procedures beginning with an examination of the external parts of the VCR, then narrowing the view to gears, springs, pulleys, belts, and other mechanical parts. This book also shows how to troubleshoot tuner/demodulator circuits, audio and video circuits, display controls, servo systems, video heads, TV/VCR combination models, and more.

 This book also contains nine VCR case studies, each focusing on a particular model of VCR with a specific problem. The case studies guide you through the repair from start to finish, using written instruction, helpful photographs, and Howard W. Sams' own *VCRfacts*® schematics.

Howard W. Sams Computer Monitor Troubleshooting & Repair

Joe Desposito & Kevin Garabedian

 Computer Monitor Troubleshooting & Repair makes it easier for any technician, hobbyist or computer owner to successfully repair dysfunctional monitors. Learn the basics of computer monitors with chapters on tools and test equipment, monitor types, special procedures, how to find a problem and how to repair faults in the CRT. Other chapters show how to troubleshoot circuits such as power supply, high voltage, vertical, sync and video.

 This book also contains six case studies which focus on a specific model of computer monitor. Using carefully written instructions and helpful photographs, the case studies guide you through the repair of a particular problem from start to finish. The problems addressed include a completely dead monitor, dysfunctional horizontal width control, bad resistors, dim display and more.

Video Technology
184 pages s Paperback s 8-1/2 x 11"
ISBN: 0-7906-1102-3 s Sams: 61102
$29.95 s March 1997

Troubleshooting & Repair
308 pages + Paperback + 8-1/2 x 11"
ISBN: 0-7906-1100-7 + Sams: 61100
$29.95 + July 1997

CALL 1-800-428-7267 TODAY FOR THE NAME OF YOUR NEAREST PROMPT PUBLICATIONS DISTRIBUTOR